# TERROR
# in the Skies

# TERROR
# in the Skies

## The Inside Story of the World's Worst Air Crashes

### by David Grayson

Published by
the Paperback Division of
W.H. ALLEN & Co Plc

A Star Book
Published by
the Paperback Division of
W. H. Allen & Co Plc
Sekforde House
175/9 St John Street
London EC1V 4LL

Printed and bound in Great Britain by
Courier International Ltd, Tiptree, Essex

ISBN 0 352 32519 4

Dedicated to my wife, Julie, who, throughout the years, has cheerfully traveled by air with me from coast to coast, as well as to Hawaii, Canada, Mexico, Bermuda, Puerto Rico, St. Croix, St. Martin, the Dominican Republic, Barbados, the Middle East and Europe—but, until our youngest reached the age of 21, *never* on the same airplane.

We now fly together.

# Contents

# Foreword

There appears to be a fascination on the part of the public and the media with any form of airplane crash. Especially when one involves a modern jet airliner resulting in a number of casualties.

For days—and sometimes weeks following an accident, newspapers, television and radio stations will devote an inordinate amount of space and time describing the accident, trying to explain how it occurred, quoting experts and interviewing witnesses to the crash and survivors, if any. However, despite the extensive coverage, the inside story has yet to be told, and the public remains curious.

This book will go a long way toward satisfying this compelling curiosity. It is a true account of the most terrifying airplane accidents experienced by major airlines—including some frightening near-misses—over the last dozen years. By the end of each chapter, you will know exactly what took place behind the scenes to cause each crash—information that was rarely made available to the general public—the inside story.

Each accident is followed through from takeoff to the termination of the flight. And, to give you the opportunity to know exactly what is going on, you will be taken along as an observer and participant on each and every one of these flights.

Part of the time, you will be seated in the cockpit as a member of the flight crew. There, you will be faced with the same dilemma as the other crew members, as they try to cope with the unexpected problems thrust upon them. You will participate in the discussions and decision-making taking place in the cockpit, and eavesdrop on the radio transmissions between your aircraft and the controllers.

At other times, you will give up your place in the cockpit as a crew member and become a passenger occupying a seat in the cabin of the airplane. In that position, you will experience the frightening and helpless sensations all passengers feel when they realize their aircraft is in danger of crashing.

You will also assume, at different times, other roles; that of a survivor, a member of a rescue team arriving on the scene, and even an eyewitness describing the accident.

By the time you have completed all of your nerve-wracking flights, you will know the true inside story of what you had only previously heard on radio or television, and seen monopolizing the front pages of the newspapers. Information, incidentally, that was, in many cases in the early stages, inaccurate or incomplete and usually never correctly reported until many months after the accident.

As we have said before, everything that you will be reading is true. The information was obtained from U.S. Government publications and newspaper reports. We gratefully express our appreciation to the many staff members of the National Transportation Safety Bureau, the Federal Aviation Administration and assorted libraries for their gracious and willing assistance in furnishing vital information and publications on request.

Now, it is time to start our takeoff. If you have a strong enough stomach to take what is coming, we suggest that you take a deep breath, try to relax, and make yourself comfortable. Be sure to move your seat to an upright position and place both feet firmly on the floor. Are you ready now? If so, fasten your seatbelt—and read on.

# TERROR
## in the Skies

# Chapter 1

# Tires Blow Out on Takeoff

On March 1, 1978, Continental Airlines, Flight 603, was scheduled to leave from Los Angeles Airport for a flight to Honolulu. On board the wide-body DC-10 aircraft were 186 passengers, three flight crew members and 11 flight attendants, which came to a total of 200 persons.

It was about 9:00 AM and most of the passengers aboard the flight were vacationers looking forward to their trip to exotic Hawaii. A number of them were part of a tour group arranged by the American Association of Retired Persons and the National Retired Teachers Association.

The DC-10 jumbo jet received permission to leave the terminal gate and was cleared to taxi to runway 6R. Although the runway was wet, it had no pools of water on it.

At 9:22 AM, Flight 603 was cleared to taxi into position on the runway and hold. A minute later, the flight was given permission to take off and power was applied to the three powerful engines to get the super-heavy jet aircraft moving down the runway. At the controls was one of the airline's most experienced pilots. He had been employed by Continental Air Lines for more than 30 years and had accumulated approximately 30,000 total flight hours, many of which were in the DC-10 aircraft.

The flight crew reported that acceleration was normal and that all engine instruments were in the normal range for takeoff. Although runway 6R was 10,285 feet long, it was expected that the heavy jumbo aircraft would use every bit of the runway. Stated the veteran First Officer (co-pilot) who had been employed by Continental for 12 years, also with many hours of experience in the DC-10, "Acceleration was good. It is a big, overpowered airplane. That is the feeling I have always had taking this thing off."

As the airplane accelerated down the runway, the ground speed approached the V-1 speed of 156 knots (180 miles per hour). This V-1 speed is known as the decision speed.

Flight crews use the V-1 speed as a decision point during the takeoff roll. If an engine failure is recognized before the V-1 speed is reached, the pilot is trained to reject the takeoff, and he, in fact, must abort the takeoff since he cannot be assured of successfully continuing. On the other hand, if the aircraft is beyond the V-1 speed before an engine failure is recognized, the takeoff must be continued since the pilot cannot be assured of stopping the aircraft on the runway remaining.

It is important to understand that as an aircraft approaches V-1, the decision-making time available to the pilot decreases. A failure to act promptly to any problem encountered as the aircraft approaches V-1 can have catastrophic results if the problem is an engine failure, or is associated with loss of flight controls. Certainly, these possibilities are ever present in a pilot's mind when something unusual occurs at a critical time. Therefore, the dominant tendency for most pilots is to abort the takeoff when any unevaluated abnormality occurs before the V-1 speed is reached.

Getting back to Continental Air Lines Flight 603 as it was accelerating on the takeoff roll: Exactly 1.2 seconds before the DC-10 reached the critical V-1 speed, the Captain heard a "loud metallic bang--if you wanted to duplicate the sound, you would have a heavy metal piece and hit it with a large hammer and it would ring, and that is what I heard," which was followed immediately by "a kind of quivering of the plane."

What had occurred at the time, at just about the most crucial point of the takeoff roll, was that one of the four tires on the left landing gear blew out. This imposed the entire load of the axle on the sister tire, which then also blew out almost immediately. Both tires then started to disintegrate, strewing shreds of rubber on the runway behind the accelerating aircraft.

This information was then not available to the Captain. All he knew was that he was faced with the need for immediate action. He had no time in which to evaluate the significance of the loud bang and vibration if he was to successfully reject the takeoff. He acted. "Abort!" he screamed.

A rejected takeoff was begun immediately. The Captain applied full brake pressure while simultaneously bringing the thrust levers back to idle power. Reverse thrust levers were actuated and full reverse thrust was used. The flight data recorder indicated that the Captain's reactions were remarkably quick in that the engines' forward thrust began to be reduced in less than one-half a second after V-1 speed.

The flight crew thought the DC-10 could be stopped in time, but then the aircraft began to vibrate. By now, another tire on the same left gear of the aircraft had also blown out leaving only one of the four tires on the left main gear intact. This condition, together with the wet runway, considerably reduced the braking capability of the jumbo aircraft.

The aircraft veered to the left of the runway centerline and, at first, appeared to the flight crew to be decelerating normally. However, with about 2,000 feet of runway remaining, the flight crew became aware that the rate of deceleration had decreased, and they now realized that the DC-10 would not be able to stop on the runway surface.

The Captain said, "The vibration was increasing very, very much. I mean it was just getting wild. In fact, I didn't feel like I was even sitting in the seat—I was being bounced off it." He added, "With this vibration, it seemed like we just started floating, and nothing was stopping us."

This buffeting-about was confirmed by the First Officer who stated that the vibration "got violent. The [control] yoke is out of my grasp and I was thrashing around. I had to clutch myself in because I was flailing against the seat. It got to the point where I could not keep my feet on the pedals. I have got bumps and bruises on my shins from being slammed up into the underside of the instrumental panel. The seat belt worked—that kept me in there."

Despite the violent vibration, the flight crew encountered no problems with directional control of the aircraft. The Captain maintained maximum brake pedal force and full reverse thrust as he steered the aircraft to the right of the runway centerline in an effort "to go beside the stanchions holding the runway lights." The stanchions, located immediately off the departure end of the runway, are constructed of heavy steel. These inflexible supports for the approach lights would have caused a thoroughly disastrous ending of Flight 603 if the captain had not been able to steer his aircraft successfully around them.

As the DC-10, spewing sparks from the three bald, scraping wheels, slid by the departure end of the runway, the Captain became aware of another danger. "There was a parking lot up there with a lot of cars so I felt, well, I better steer around behind the lights."

Said the First Officer about the Captain, "I am talking him through this darn thing . . . 'Fight, fight, stay there . . .' Beyond that runway, there is an overrun area and I could see all these cars out there. He is on the brakes and reverse [engine thrust]. We are vibrating—banging, banging, banging—all of a sudden, the jet rolls off the runway and the left landing gear collapses."

In a hearing before the National Transportation Safety Board, the Second Officer, who was the flight engineer as well as a pilot, gave his recollection of what happened. His testimony, although enlightening, was somewhat amusing in the way he expressed himself. "I was not alarmed. Things were not going, you know, like on a Sunday drive, but I wasn't real nervous about it. I knew something was wrong. Everybody knew something was wrong.

"I thought at that time, we are going to stop. It is screwing up my Honolulu trip, was my thinking. Now I am going to have to go in and get some new tires, and then we will go.

"I then had the thought, maybe we are not going to get it slowed down to make a nice right-hand 90 degree turn and go on our way. It was not going to work out that good.

"[As] it went off [the runway] it just swerved. Right there it got violent, and the instruments were blurred. It bounced to a stop and you have the oh shit, the normal oh shit thoughts. Here, look at this mess.

"Just then, the smoke curled around the cockpit, followed by a ball of fire, and I thought, well, this is bad news. I reached up and grabbed the P.A. [public address microphone] and said, 'Easy, Victor, Easy'—you know, [the signal for] the get out—the evacuation thing, a couple of times."

The DC-10 had slithered off the runway onto a thin layer of asphalt which broke through under the heavy load of the jumbo jet. The left main landing gear then sheared off causing the aircraft to drop onto its left wing, scraping the pavement as it continued to slide along the surface. The shower of sparks then ignited the jet fuel spilling out of the ruptured wing tanks. Flight 603 finally stopped 664 feet past the end and 40 feet to the right of the runway.

A maintenance supervisor arriving on the scene said, "The fuel was ankle deep and flames were shooting higher than the airplane. We were helping people out and others were jumping. . . ."

Another person on the scene stated, "There was a sea of fire. Those poor bastards had to jump into the fire."

When the airport firemen arrived, they saw passengers jumping or sliding down the emergency chutes out of the plane. "Everybody was running for their lives," said the fire department captain. He also related that some dazed passengers were sitting or lying on the ground while the fire burned furiously nearby.

The fire chief later reported that all the exit chutes were deployed but some of them failed to inflate. He felt that many of the passengers who tried to slide down the limp chutes sustained injuries doing so. Some received rope burns from sliding down the emergency rope which hung from the First Officer's side window. Also, many passengers in the rear of the plane were injured when they were forced to jump the long distance to the ground to avoid the invading flames.

Some passengers were too frightened to follow instructions issued by the crew members.

A couple, a man and his wife in their seventies, exited through an emergency door over the left wing. Then, despite shouted warnings from a stewardess not to use that exit, the two slid down a chute off the forward edge of the wing—right into the flames. At that point, there was no escape—it was impossible for the pair to get to safety.

Recalled a survivor, "They just jumped into hell . . . like the one they preached when I was a kid."

Another passenger was also shaken by the memory of that incident.

"There was all this confusion . . . and then a rush for the emergency exit. Right then I heard one of the stewardesses saying, 'Don't do that! Please don't do that! Don't open that door! Not that one!'

"She was talking about the exit on the left side. She could see what had happened—that there was nothing but fire out there and she knew it was wrong to go that way. She was fighting her way toward that door . . . to guard it, keep people away, I guess.

"But then there was a pop, and the left door was open. And people were jumping . . ." The horror of that scene was reflected in his face.

Despite the imminent danger of an explosion, knowing their lives were in jeopardy, all the crew members heroically remained with the burning aircraft to calm and assist the milling, confused passengers in the evacuation.

The Captain, in his testimony before the Safety Board, recalled that almost immediately after the aircraft came to a stop, a flight attendant stuck her head in the cockpit and informed the flight crew that the aircraft was on fire. While the First and Second Officers were going through the standard emergency procedures, which included shutting off the fuel flow, he entered the cabin.

"I could see [emergency doors] open . . . people were crowding in, not pushing . . . everybody was handling it right," he said. "People were sliding down the [chutes] . . . it seemed organized . . . [but] a little slow. In this [front] section, it was mostly little grandmas coming off . . . I would catch them at the knees and set them out there . . . I was getting them out there pretty fast," he recalled.

The Captain then jumped out of the aircraft and said he saw

"three or four people lying in the mud, with water, kerosene and foam mixed up, and they were all old granny-type people, old people, lying on one another down there, and the fire department truck was sitting back there 20 feet . . . squirting the foam in there on these people. I started helping them out.

"The fire had developed to a point," he continued, "[where] it sounded like a huge torch going . . . it was really an intense fire . . . It just looked like that with this fire, the plane would just have to blow at any time. I got the truck to come up further and said, 'Just keep the foam going.'" The fire truck drove up a little and continued spraying foam. The Captain then told a fireman, "Just keep it coming on me," and as "he kept the foam coming on me, [I] kept hauling these people out."

Added the Captain, "If they hadn't been so elderly, it would have been no problem . . . but they just seemed immobile. They would sit there and you would have to pick them up, every one."

The First Officer was the next person before the Safety Board. "I went out my partially open window," he said, "down the rope, ran to the bottom of the slide, pulled it out to give it some slope, yelling, 'Let's go. Come on. Let's go.' They came out, two's and three's.

"And so it was just a matter of out as fast as possible. Here is all this fire coming this way under the airplane, big, black smoke all over the place, like blocking out the sky. I thought we were dead. This thing is going to blow up.

"In fact, I can remember thinking I am gone. Back to the slide to get as many people out before it goes. Then just kept talking them down."

Finally, said the First Officer, the last three on the airplane (two of them hostesses), came down the rope. He stated, "I caught her [the last one down] and I thought, 'Now she can blow.' I walked away."

The Second Officer also believed an explosion was imminent but that didn't stop him from rescuing passengers.

"I heard one of the slides explode. Bang; something burnt or exploded. So I went down the rope. Got on the ground. The co-pilot was down there holding the slide, him and another Continental Captain. The Captain was in back of the wing somewhere, doing something and the co-pilot and I were hauling people over and setting them down.

"And there was fuel about that deep all over . . . it was sort of dangerous, you know. My only thought was, well, I got one more trip in after one more person, then the damn thing will blow up. Each time I would get somebody out I would think, well, I got one more trip coming and I would go back in there and it didn't blow up. I don't know how many times I did that, or why I even did it, because it was really burning."

While the crew members were catching and cushioning the jumps of the passengers, one young girl in her twenties, holding a very young baby, walked up to the Second Officer and said in a frightened voice, "I fell on the baby. I don't know if he is all right."

The Second Officer pulled away the covering from the baby's head where they could both see blood on the face. The mother started screaming until the flight crew member said, "The blood is off you." He then pointed out the fact that the mother had a little cut on her eye which had dripped on the baby's face. The Second Officer said, "She then was happy about it. I took my coat, covered the woman and went back to doing what I was doing."

Flight attendants recalled that many passengers were scared to use the slides. Stated one, "I started screaming for the people to get out and there was a slight reluctance. . . . I just started kicking and shoving and hitting and screaming as loud as I could scream . . . trying to impress them to move faster, because I could see the fire coming closer and closer."

Passengers had their own stories to tell. One said, "I lost my purse, and I lost all my money, and I lost my traveler's checks, and I probably lost all my baggage that was aboard, too. So what? I didn't lose my life. I'm lucky!"

Another recalled that she prayed a short, frightened prayer when the aircraft bumped to a stop—not at all like the one voiced by her 15-year-old daughter.

"She's a real Christian," said the mother. "I heard her say, 'In the name of Jesus Christ, I rebuke you, Satan, from this plane.' Then the fire began and I screamed, 'Let's get out of here!'"

"My mother has a heart condition," explained the teenager. "She was excited. But something must have worked, you know. . . ."

Of the 200 persons on board, four passengers were fatally injured. Three crew members and 28 passengers were seriously hurt with 11 crew members and 156 passengers escaping with either minor or no injuries at all.

One firefighter was seriously hurt and nine firefighters were injured slightly while extinguishing the fire.

The National Transportation Safety Board highly commended the Los Angeles Fire Department. They stated that the quick response "prevented greater loss of life and lessened injuries to evacuees." This was possible because of the location of fire stations at the midpoint of LAX's two major runway complexes. Upon hearing the noise of the tire blowouts, firemen were on their way even before the jet went off the runway. Foam was sprayed immediately, driving back the flames and keeping escape lanes open for the evacuating passengers. The Safety Board suggested that other major airports, which may be having difficulty with their emergency response times, follow the example set at Los Angeles International Airport.

The Board also commended the crew members. "The success of the emergency evacuation of the passengers, most of whom were elderly, was the direct result of the efforts of the entire flight crew and cabin crew and that of a Continental B-727 Captain who was onboard as a passenger. . . ."

This accident was Continental's first fatal one in 20 years. As a matter of fact, it was the first time a commercial airliner accident had killed anyone on the Los Angeles Airport grounds. (There had been a couple of fatal accidents in the past with airliners that had crashed into the ocean west of the airport.)

According to a spokesman for McDonnell Douglas, up to the day of this accident, the DC-10's safety record for the preceding four years had been excellent. It is one of the most widely used of the commercial jumbo jets, with 247 of them flying for 35 different airlines around the world. Unfortunately, this excellent safety record was thoroughly destroyed the following year, in 1979, when a DC-10 crashed shortly after takeoff from the

Chicago-O'Hare Airport with a loss of 273 lives.

In the investigation, it was revealed that the DC-10 took off on runway 6R, which was 10,285 feet long. There are two other runways that are 12,000 feet long at the same airport, 25R (right) and 25L (left). Since Flight 603 skidded to a stop only 664 feet past the end of runway 6R, obviously, the aircraft could have safely come to rest on the runway had it been able to use one of the longer ones, 25R or 25L.

However, the problem is that the longer runways are restricted to aircraft not exceeding 325,000 pounds because of runway overpass strength limitations. Flight 603, because of its gross weight of 430,000 pounds, therefore only had runway 6R available to it for takeoff. The Safety Board urged, "The responsible authorities to . . . (expedite a project) to make longer, safer runways available to heavier aircraft as Los Angeles International Airport."

The Safety Board made another important statement. They felt that the current training for rejecting a takeoff, which is based solely on engine failure, should be modified to also take into consideration wet runways, tire failures, and the deposit of rubber on the surface of a runway caused by the tires of aircraft landing in the opposite direction. (There was a heavy deposit of rubber from tire contact on landings on the runway used by Flight 603, further reducing runway friction and braking capability.)

Another recommendation was that the Federal Aviation Administration assess current tire rating criteria. Tire blowouts due to heat generated by high taxi speeds, long taxi distances and excessive use of brakes have become more common and, potentially, more dangerous.

The DC-10 had experienced tire problems in the past. A Continental Air Lines spokesman revealed that the DC-10's operated by their airline suffered 29 tire failures in the last two years. Nevertheless, he said that none of these resulted in accidents and that the company DC-10s made over 100,000 takeoffs and landings in that time.

However, an official from the Federal Aviation Administration stated that the FAA had been sufficiently concerned that it was preparing a directive that would make mandatory the installation of shields in the wheel wells of all DC-10 aircraft to protect hydraulic systems from possible damage by threads thrown off by blowouts.

There are a number of interesting sidelights to the Continental DC-10 accident worth mentioning. One of them had to do with the age of the Captain.

According to FAA regulations, airline captains must turn over the controls to younger pilots at the age of sixty. The mandatory retirement rule exists because "the risk of a pilot suffering an incapacitating in-flight event gets too high" after that, according to a spokesman for the Federal agency. An over-age pilot, however, can continue flying as a flight engineer, with a cut in pay and seniority, if he so desires and a position is available.

The Captain of Flight 603 was 59 years old and his flight to Hawaii was his last official one before retiring. According to tradition observed by many pilots, his wife accompanied him on this last flight, as a passenger.

The Captain observed how ironic it was that he had been flying for Continental Air Lines for 32 years and had logged more than 30,000 hours without experiencing one accident. Up to the day of his retirement flight, he stated, "I had never scratched a passenger and never dinged an aircraft."

As for his wife, she had exited the flaming aircraft without any assistance from her husband. She was unhurt. "She's a pretty versatile gal," he said. "She knows how to take care of herself."

Finally, a note of irony. When a question was raised at the hearings as to how the number of dead and injured could have been reduced, the Captain stated that, although the plight of the passengers in the aircraft being consumed by fire seemed precarious at the time, in the final analysis, all of the passengers and crew members would have been much better off remaining in the aircraft while the firemen extinguished the flames. "Hindsight being wonderful," he testified, "knowing they got the fire out, if they had all been sitting in the airplane, they would have been all right, thanks to the firemen."

# Chapter 2

# In a Hurry

In the previous chapter, the flight crew thought they were doing the right thing by ordering the emergency evacuation when the flaming aircraft screeched to a stop. Unfortunately, as a result of what initially appeared to be a sound decision, four people perished and 29 other Hawaiian-bound passengers were seriously injured.

We now turn to another vacationing group, also traveling on a jumbo airliner, where an unsound decision by the Captain (according to the investigating authorities) was primarily responsible for the resulting catastrophe. The events leading up to this major accident were as follows:

On March 27, 1977, a bomb exploded in the passenger terminal at Las Palmas airport in the Canary Islands, Spain, temporarily closing the airport to all traffic. All aircraft with a Las Palmas destination were diverted to the nearby airport on the island of Tenerife, less than 70 miles away.

One of the airliners re-routed to Tenerife awaiting the reopening of Las Palmas was a Pan Am 747 originating from Los Angeles, California, with 16 crew members and 378 passengers, mostly vacationers. Also at Tenerife, awaiting permission to take off and land at its primary Las Palmas destination, was another giant four-engine 747, a KLM jet from Amsterdam with 14 in crew and 234 passengers.

The delay, so far, had lasted more than two hours. Although everyone was impatient to get going, the KLM Captain was especially anxious to depart. His crew had already flown sufficient hours so that any further delay could put them in the position of possible exceeding their limit of continuous crew time for that tour of duty. Since these limits are strictly enforced, this would compel them, with their passengers, to remain (most unhappily) overnight in Tenerife.

A short time later, to the applause of the passengers, the stranded aircraft received the welcome news that Las Palmas was now open for landings. At about 5:00 PM (local time), the KLM 747 received radio permission to taxi out on the runway (Tenerife only had one runway) to the very end where it would make a backtrack (a 180 degree turn) and prepare for a takeoff in the direction from which it had just taxied. This one runway, normally used only for takeoffs and landings, had to be utilized for taxiing since all the taxiways running parallel to the runway were clogged with aircraft diverted from Las Palmas.

Immediately after the KLM started taxiing down the runway, the 747 Pan Am received permission to taxi behind the KLM—but with the instructions to turn off at one of the exits midway down the runway (actually, the third one). This was necessary to move the Pan Am 747 out of the way of the KLM 747, which it was following, since the latter would be turning around at the end and taking off in the opposite direction on the same runway.

It is important to note here that, when the planes originally landed, the weather was sunny and clear. Now, however, low-lying clouds had rolled in and visibility was limited to a few hundred yards. Therefore, the controller in the tower could not see the planes nor could either of the 747 pilots see the other aircraft.

We therefore have a situation in which the 747 KLM, in this patchy fog, is taxiing down to the end of the runway with instructions to turn around at the end and await permission to take off in the direction it just came from. Taxiing behind the KLM is the

Pan Am 747 (also with very limited visibility) with orders to turn off at the third runway exit (a little past the halfway point on the runway) so it can be safely off the runway when the KLM 747 receives permission to take off.

It is now time to hear the actual conversations taking place in the cockpit between crew members among themselves and with the controller in the tower. This information comes from tapes taken from both aircrafts' cockpit recorders and transmission tapes furnished by the Tenerife Control Tower. Some of the conversations have been deleted to maintain continuity. Also some words were inserted [in brackets] for purposes of clarification.

### TIME 5:02:03(seconds) PM

| | |
|---|---|
| Pan Am Co-Pilot: | We were instructed to contact you and also to taxi down the runway. Is that correct? |
| Tower: | Affirmative. Taxi onto the runway and leave the runway third [exit], third to the left. |
| Pan Am Co-Pilot: | Third to the left, okay. |
| Pan Am Captain: | Third, he said. |
| Tower: | [Th . . .]ird one to your left. |
| Pan Am Captain: | I think he said first. |
| Pan Am Co-Pilot: | I'll ask him again. |
| Pan Am Captain: | [To another crew member] What really happened over there [in Las Palmas] today? |
| Pan Am Unknown Voice: | They put a bomb in the terminal, sir, right where the check-in counters are. |
| Pan Am Captain: | Well, we asked them if we could hold [wait out the delay by flying a holding pattern] and I guess you got the word, we landed here. |

### TIME 5:02:49 PM

| | |
|---|---|
| Tower: | KLM 4805, how many taxiways did you pass? |

| | |
|---|---|
| KLM: | I think we just passed Charlie 4 [the fourth and last taxiway] now. |
| Tower: | Okay, at the end of the runway make 180 [degree turn] and report [when] ready for ATC [Air Traffic Control] clearance. |

## TIME 5:03:09 PM

| | |
|---|---|
| Pan Am Co-Pilot: | The first one [taxiway] is a 90-degree turn [to get off the runway]. |
| Pan Am Captain: | Yeah, okay. |
| Pan Am Co-Pilot: | Must be the third. I'll ask him again. |
| Pan Am Captain: | Okay, we could probably go in, it's . . . |
| Pan Am Co-Pilot: | You gotta make a 90 degree turn . . . |
| Pan Am Captain: | Yeah. |
| Pan Am Co-Pilot: | . . . 90 degree turn to get around this. This one down here [is] a 45 [degree turn to exit off the runway]. |

Obviously, there was confusion in the Pan Am cockpit.

## TIME 5:03:29 PM

| | |
|---|---|
| Pan Am Co-Pilot: | [To tower], would you confirm that you want the [Pan Am] Clipper 1736 to turn left at the thirrrrrrd intersection? |
| Tower: | The third one, sir; one, two, three; third, third one. |
| Pan Am Captain: | Good. |
| Pan Am Co-Pilot: | Very good, thank you. |
| Pan Am Captain: | That's what we need, the third one. |
| Pan Am Engineer: | Uno, dos, tres. |
| Pan Am Captain: | Uno, dos, tres. |
| Pan Am Engineer: | Tres, sí. We'll make it yet. |
| Tower: | Pan Am, report [when] leaving the runway [via the third taxiway exit]. |

The Pan Am 747 crew acknowledged the most recent instructions and while looking for the third turn-off from the runway, busied themselves with the check list and reviewed departure procedures. So far, they had passed the first two exit turn-offs.

By now, the KLM crew had reached the end of the runway, turned around and was making takeoff preparations.

### TIME 5:04:58 PM

| | |
|---|---|
| Tower: | KLM and [Pan Am] Clipper 1736. For [both] your information, the centerline lighting is out of service. |
| KLM: | I copied that [message acknowledged]. |
| Pan Am: | Clipper 1736 [message acknowledged]. |

### TIME 5:05:22 PM

Pan Am 747 continues taxiing along the runway looking for the third taxiway exit.

| | |
|---|---|
| Pan Am Captain: | That's two. |
| Pan Am Engineer: | Yeah, that's a 45 [degree turn-off] there. |
| Pan Am Captain: | Yeah. |
| Pan Am Co-Pilot: | That's this one right here. |
| Pan Am Engineer: | Okay, next one is almost a 45 [degree turnoff]. |
| Pan Am Captain: | But it goes . . . ahead. I think (it's) gonna put us on [the] taxiway. |
| Pan Am Engineer: | Yeah, just a little bit, yeah. |
| Pan Am Co-Pilot: | Maybe he counts these [for] three. |

The Pan Am crew remained confused. They missed the third turnoff and continued taxiing down the runway searching for their designated exit.

At about the same time, the KLM 747 crew had completed their cockpit procedures and were apparently awaiting permission to take off. However, the digital flight data recorder on the KLM indicated a slight forward movement of the aircraft due to opening of the throttle at:

### TIME 5:05:41 PM

| KLM Co-Pilot: | [To KLM Captain] Wait a minute. We don't have ATC [Air Traffic Control] clearance. |
| KLM Captain: | No, I know that. Go ahead, ask [for it]. |
| KLM Co-Pilot: | [To Tower] Ah, the KLM 4805 is now ready for takeoff and we're waiting for our ATC clearance. (This communication was also heard in the Pan Am cockpit since both aircraft are tuned to the same tower frequency.) |
| Tower: | KLM, you are cleared to . . . (Instructions are given as to the turns to be made after takeoff, the altitude to climb to and the heading to take towards Las Palmas radio station.) |

As a matter of procedure, the receipt of an Air Traffic Control clearance does *not* give an aircraft permission to take off. It has only received the routing and altitude the plane will fly when it does depart. The pilot must first receive another radio command that he is cleared to take off *before* he is allowed to start his takeoff.

## TIME 5:06:09 PM to 5:06:17 PM

During this time period, the co-pilot of the KLM confirms the ATC clearance by reading back the clearance instructions to the tower. However, while the co-pilot is talking to the tower, the flight data and Cockpit Voice Recorders indicated (in that same eight-second time interval of above) that the KLM Captain released the brakes, told his crew "Let's go," advanced the throttles for forward thrust and started his takeoff before the co-pilot finished his readback. (A note of interest here is that the 50-year-old KLM Captain was his airline's Chief-of-Pilot Training, a position of great respect and authority. The other KLM crew members were aware that they were flying with one of the pilots of greatest prestige in the company and were not likely to challenge his authority.)

TIME 5:06:18 PM                    TIME 5:06:18 PM

The KLM co-pilot had just finished his readback and as his aircraft starts to move, now hurriedly says:

KLM Co-Pilot: [To Tower] We are now at takeoff. (The normal interpretation is that he is at the takeoff position and still needs radio permission to actually start his ground run.)

Pan Am: [To Tower] We are still taxiing down the runway, the Clipper 1736. (This Pan Am transmission partially blocked out the tower's message to KLM.)

Tower: Okay [KLM]. Stand by for takeoff. I will call you.

(Due to the blocked transmission, the only word from the tower that the KLM heard clearly was "Okay." KLM did not hear, "Stand by for takeoff. I will call you.")

TIME 5:06:25 PM

Tower:          Pan Am 1736, report runway clear [when you are clear of the runway].
Pan Am:         Okay, we'll report when we're clear.
Tower:          Thank you.

(These transmissions from Pan Am to the tower were audible in the cockpit of the KLM which had already started its takeoff run.)

TIME 5:06:32 PM                    TIME 5:06:32 PM

KLM Engineer: [In the KLM cockpit] Is he [the Pan Am] not clear, then?

KLM Captain: What do you say?

Pan Am Captain: [In Pan Am cockpit] Let's get the hell right out of here.

KLM Engineer: Is he not clear,     Pan Am Co-Pilot: Yeah, he's
   that Pan American?          anxious, isn't he?
KLM Captain: Oh yes!               Pan Am Engineer: Now he's in
                                 a rush.

### TIME 5:06:40 PM

Pan Am Pilot:         There he is. . . . Look at him! Goddamn!
                          That—that son-of-a-bitch is coming!!
Pan Am Co-Pilot:   Get off! Get OFF!! GET OFF!!!

The Pan Am 747 applied full throttle as it frantically tried to turn off the runway at a 45-degree angle relative to the center of the runway. The KLM 747 crew, finally aware of the other 747, desperately attempted to pull their aircraft into the air up and over the Pan Am airplane. The KLM actually managed to get entirely airborne; however, it was not enough. Although the nose and front landing gear of the KLM just managed to clear the pan Am, the main landing gear smashed against the Pan Am fuselage.

Each aircraft weighed approximately 700,000 pounds!

### TIME OF IMPACT 5:06:50 PM

The KLM, after destroying the top of the Pan Am fuselage, desperately attempted to gain altitude. Like a staggering bird, the 747 struggled and strained in an effort to remain airborne—but to no avail. After 500 feet of flight, it gave up, fell back to the runway and slid an additional 1,000 feet before coming to a complete stop. The aircraft then caught fire suddenly and with such force that emergency evacuation operations could not be employed.

Although the body of the KLM 747 remained substantially intact, there was no sign of movement from the aircraft. The doors stayed shut. All that could be heard was the roaring and raging fire. There were no survivors.

In the Pan Am 747 aircraft, the first-class lounge disintegrated as a result of the impact. The lounge floor also gave way which meant that the crew had to jump to the first-class section and get out through a hole in the left wall. This hole was also the main escape route for the passengers located in the forward part of the aircraft.

At the center and rear of the Pan Am plane, the twisting of metal sheets of the fuselage, along with the fire which suddenly broke out, formed a kind of trap preventing forward exit of the passengers. Those who survived managed to jump to the ground through an opening on the left side or through an open door from a height of 20 feet. Despite a fire under the left wing, some of the passengers managed to escape by jumping off this wing onto the grass.

Of the 16 crew members and 378 passengers aboard the Pan Am 747, only 59 survived. All of the survivors suffered injuries.

There were unusual circumstances surrounding this disaster. The weather conditions with its fog patches prevented this accident, despite its magnitude, from being directly visible from the control tower. Actually, when the control tower heard one explosion followed by another, it was unable to pinpoint them on the airfield and also unable to determine the cause of the explosions. It was some moments later when an aircraft located on the parking apron advised the tower that it had seen a fire, without specifying the exact place or its cause. Still not being able to locate the fire, the tower alerted the fire service.

The firemen headed for a bright light through the fog. When they came closer, although they were as yet unable to see the flames, they were hit with the effects of strong heat radiation.

When there was a slight clearing, they saw for the first time that there was a plane (the KLM 747) totally enveloped in flames, the only visible part being the rudder. After they had already begun to fight the fire, a greater clearing in the fog took place and they saw a bright light further away, which they thought at first was a part of the same plane which had broken off and was also burning.

They divided up the fire trucks and, on approaching what they thought was only an additional segment of the same fire, discovered a second plane on fire (the Pan Am 747). The firemen immediately concentrated their efforts on this second plane because the first was already totally beyond salvaging.

Meanwhile, in all this time, because of the dense clouds surrounding it, the tower was still unaware of the exact location of the fire and whether one or two planes had been involved in the accident. It was some time before the confusion dissipated and the account of what actually occurred started to emerge.

The most horrifying fact was the size of the death and injury toll. The total number was the largest ever recorded in an aircraft accident. The figures for each aircraft broke down as follows:

### KLM 747

|          | CREW | PASSENGERS | TOTAL |
| -------- | ---- | ---------- | ----- |
| FATAL    | 14   | 234        | 248   |
| INJURIES | 0    | 0          | 0     |

### PAN AM 747

|          | CREW | PASSENGERS | TOTAL |
| -------- | ---- | ---------- | ----- |
| Fatal    | 9    | 317        | 326   |
| Injuries | 7    | 61*        | 68*   |

*Of the 61 passengers injured, nine subsequently died as a result of the injuries received.

Adding the casualty list of both 747 aircrafts comes to a mind-boggling total of 583 killed and 59 injured as a result of one aircraft impacting another. These numbers are especially hard to assimilate when we realize that this catastrophe occurred when the wheels on the main landing gear of one airplane *just did not quite clear* the fuselage of the other on takeoff.

Five hundred eighty-three people died. Not even taking into consideration the considerable number of injured, how can we possibly grasp the enormity of that number of deaths from one accident? Five hundred eighty-three people.

In an attempt to come to grips with what could be viewed by some as merely a statistic, consider the following comparisons: 583 deaths comes to more individuals than all of the players on 64 baseball teams; or, as many athletes on 53 complete football teams; or, more than all the teammates on 116 basketball teams. In fact, the number is about the size of an entire school body of a small-town high school, or the entire seating capacity of an average-sized movie theatre house. Obviously, a number too large to easily comprehend.

What about the ramifications of these numbers? How about the relatives of these unfortunate victims? The spouses, children, parents, sisters and brothers? And even boyfriends, girlfriends, fiancées (and fiancés)? How about employers or key employees of the deceased? Or business partners? And, besides shattered family and business relationships, consider the loss by good friends, neighbors, organizations, charities, etc. The almost endless circle of people significantly affected runs into the many, many thousands.

Three hundred fifty tons of aircraft hurtling down the runway at 175 miles per hour toward another 350-ton monster. The Pan Am cockpit recorder indicated that the crew saw the KLM 747 approximately nine seconds before the impact. What was going through the minds of both crew members in those few seconds? Could the KLM Captain be saying, "Dear God, I was too hasty—I never expected the Pan Am to still be on the runway!" How about the KLM co-pilot or engineer, "I don't care if you are the Captain. I told you that we must not take off until absolutely certain that the Pan Am is clear of the runway!" What were the thoughts of any of those passengers on both aircraft who were able to see the other 747 only seconds before the impending crash?

Naturally, there was an extensive inquiry. Investigators from Spain, the Netherlands and the United States participated in trying to determine the cause of this tragic accident.

In addition to all other phases of the investigation, they listened to the tapes recovered from the cockpit recorders of the two aircraft as well as the tower tapes. The most significant evidence came from the cockpit recorder of the KLM.

Of course, listening to the tapes was only part of the investigation. It was only after every aspect of the accident was thoroughly examined that a number of conclusions were reached. Following is the Official Report of Conclusions by the authorities:

The fundamental cause of the accident was the fact that the KLM Captain:
1.  Took off without clearance.
2.  Did not obey the "stand by for takeoff" from the tower.

3. Did not interrupt takeoff on learning that the Pan Am was still on the runway.

4. In reply to the Flight Engineer's query as to whether the Pan Am had already left the runway, replied emphatically in the affirmative.

Now how is it possible that a pilot with the technical capacity and experience of the KLM Captain, whose state of mind during the stopover at Tenerife seemed perfectly normal and correct, was able, a few minutes later, to commit a basic error in spite of all the warnings repeatedly addressed to him?

An explanation may be found in a series of factors which possibly contributed to the occurrence of the accident:

1. A growing feeling of tension as the problems for the Captain continue to accumulate. He knew that, on account of the strictness in the Netherlands regarding the rules on the limitations of duty time (he may be prosecuted under the law if he exceeds the limits), if he did not take off within a relatively short space of time, he might have to interrupt the flight—with the consequent disruption for his company and inconvenience for the passengers.

2. The fact that two transmissions took place at the same time. The "Stand by for takeoff . . . I will call you" from the tower coincided with Pan Am's "We are still taxiing down the runway," which meant that the transmission was not received with all the clarity that might have been desired.

3. Inadequate language. When the KLM co-pilot repeated the ATC clearance, he ended with the words, "We are now at takeoff." The controller, who had not been asked for takeoff clearance and who consequently had not granted it, did not understand that they were taking off. The O.K. from the tower, which preceded the "Stand by for takeoff" was likewise incorrect—although irrelevant in this case because takeoff had already started about six and a half seconds before.

4. The fact that the Pan Am had not left the runway at the third intersection. This plane should, in fact, have consulted with the tower if it had any doubts, and this it did not do. However, this was not very relevant either since Pan Am never reported the runway clear, but to the contrary, twice advised that it was

taxiing on it.

5.   Unusual traffic congestion which obliged the tower to instruct aircraft to taxi on the active runway—a procedure which can be potentially dangerous.

That was the official report released by the investigating authorities. However, there are four additional items of interest worth mentioning which involved making a crucial decision:

1.   The KLM flight was part of a charter series operated on behalf of a Holland Travel group. Accompanying this group was a company travel guide. This lady decided to remain on Tenerife and did not return on board the KLM for the fatal flight to Las Palmas. Can we believe that at one time in every person's life, he (or she) will be faced with a decision on a course of action that will have a major impact on his future?

2.   The Pan Am flight allowed two additional company employees to board the plane at Tenerife and sit on the cockpit jumpseats for the flight to Las Palmas. These two Pan Am employees escaped death but were injured in the accident. Another decision by two people with a severe impact on their lives.

3.   The Tenerife airport itself is set in a sort of hollow between mountains. Therefore, on account of its altitude and location, the airport has distinctive weather conditions with frequent presence of low-flying clouds affecting visibility. Visibility before and during the accident was both minimal and quite variable. There was a threat of even a further reduction of the already precarious visibility. Faced with this threat, the way to meet it was either by taking off as soon as possible or refraining from taking off—a possibility which certainly must have been considered by the KLM Captain. His decision—takeoff.

4.   Before taxiing out, the KLM took an extra 30 minutes to refuel. Its tanks were filled with almost 15,000 additional gallons even though the 747 had more than enough fuel for its flight to Las Palmas. If we bear in mind that the Tenerife-Las Palmas flight is only one of about 25 minutes duration, the taking on of this additional weight of fuel leads us to suppose that the KLM Captain wished to avoid the difficulties of refueling in Las Palmas, with the resulting delay, because a great number of planes

diverted from Tenerife would be going there later. Question: How much less runway would the KLM 747 have used to become airborne if it had not taken on this extra load (almost 100,000 pounds) of fuel? Perhaps enough to completely clear the fuselage of the Pan Am 747? This decision, in the light of what happened, could have been the most crucial and devastating of them all.

Chapter 3

# Collision in the Air

Now you know the inside story of what really caused the Tenerife air catastrophe. Information that would have been almost impossible to pick up from the routine news reports.

The collision of the two 747's—one in the air and one on the ground—turned out to be the world's worst air disaster in the history of passenger airline travel.

Let us turn now to a different type of collision; one involving two aircraft in normal flight. However, in this chapter, you will observe that this accident, too, was unique. You will sense this immediately upon considering the following question:

What are the chances of 2 airplanes—in *broad daylight,* observed on the *same* radar screen, directed by the *same* controller and warned *many times* that they were on a converging course—ending up in a mid-air collision?

A million to one against it? Don't bet on it.

Let us go back to September 25, 1978, when Pacific Southwest Airlines, Flight 182, departed Los Angeles at 8:34 AM on its daily scheduled flight to San Diego. Aboard the three-engine jetliner were 128 passengers and a crew of seven.

The First Officer was at the controls of the 727 aircraft while the captain was conducting almost all of the air-to-ground communications. Also on this flight was a "hitch-hiking" company pilot occupying the forward observer seat which, together with the flight engineer, came to four pairs of professional eyes in the cockpit—more than enough to keep a safe distance from other aircraft.

On the same day and about the same time that PSA Flight 182 took off from Los Angeles, a single-engine airplane departed from a small airfield in California on an instrument training flight. A flight instructor occupied the front right seat and another pilot, who was receiving instrument training, sat at the controls in the left seat. These two pilots were the only persons on board the small Cessna airplane.

The Cessna made a short hop to Lindbergh Field, San Diego (the same destination as the PSA airliner), where two practice instrument approaches to a runway were flown. Ending its second practice approach, the small airplane began a climb-out away from the airport and, shortly after, was advised by the controller not to exceed 3500 feet in altitude.

While the small airplane was climbing, PSA Flight 182 was descending. When the airliner got down to 7000 feet, it informed the controller that it had the airport in sight. The San Diego controller then cleared Flight 182 to continue to descend so as to make a visual approach to the runway.

Flying conditions in the San Diego area that morning were excellent. The official recorded weather was 85 degrees temperature, winds calm, sky clear, with visibility of approximately 10 miles.

Just before 9:00 AM, the radar controller advised Flight 182 that there was an aircraft up ahead, to which Flight 182 responded that it would be looking for it. However, several seconds later, the controller was specific in the following message, "Additional traffic [another aircraft], 12 o'clock [straight ahead], three miles, just north of the field, northeastbound, a Cessna 172 climbing . . . out of one thousand four hundred [feet]." According to the Cockpit Voice Recorder, the co-pilot acknowledged and replied that he saw the Cessna.

Despite the co-pilot's response that the other aircraft was sighted, the radar controller again alerted the large jet, "Traffic's at 12 o'clock, three miles, out of one thousand seven hundred." This time, in addition to the co-pilot saying, "Got 'em," the Captain assured the controller, "Traffic in sight."

The controller then instructed Flight 182 to "maintain visual separation" (it was now the responsibility of the crew to keep away from other aircraft). The controller then radioed the Cessna pilot that there was "traffic at six o'clock [directly behind you], two miles, eastbound, a PSA jet inbound to Lindbergh, out of [coming down from] three thousand two hundred [feet], has you in sight." The Cessna pilot acknowledged, "Roger."

(Let us pause for a moment to reconstruct what has taken place up to now. At this point, the PSA jet was at 3,200 feet on its way down to landing at Lindbergh Field, while the small Cessna aircraft was climbing from 1,700 feet after its practice approach from that same airport. It is also extremely important to note here that, within the last eleven seconds, PSA 182 had been notified by the controller that there was "traffic at 12 o'clock [straight ahead], three miles . . ." and the Cessna had then, in turn, been alerted to "traffic at six o'clock [directly behind], two miles . . ." Thus, the pilots of both aircraft were cautioned that the larger, faster jet, on the same heading, was rapidly overtaking the smaller, slower, single-engine airplane.)

Flight 182 reported to Lindbergh tower that they were on the downwind leg for landing. The tower acknowledged the transmission and threw in one more warning to Flight 182 that there was "traffic, 12 o'clock, one mile, a Cessna."

The Captain, hearing this, became a little uneasy. He asked his co-pilot if that was the airplane for which they were supposed to be on the lookout. The co-pilot answered that it was, but that he had lost sight of it. The Captain then informed the controller that, although the PSA crew had the Cessna spotted before, they believed that it had already passed them.

However, after this message to the controller, the Cockpit Voice Recorder showed that the crew continued to discuss the whereabouts of the Cessna, with one crew member "asking" whether they were clear of the small airplane, another "supposing" they should be by now, a third "guessing" that they were, and the fourth "hoping" so. The Captain ended the discussion with an authoritative "Oh yeah, before we turned downwind, I saw him about one o'clock [a little to the right], probably behind us now." After that comment, the landing gear was ordered to be lowered and the crew busied themselves with landing preparations. (Four highly-trained flight crew members in the cockpit and not one of them was able to keep the Cessna continuously in sight.)

At 9:01:28 AM, a conflict alert warning signal was set off in the San Diego Approach Control Facility, indicating to the controller that the flight paths of PSA 182 and the Cessna had converged to a point where they were in serious danger of colliding. Despite the fact that the controller believed PSA 182's flight crew had been aware of the small airplane's position, and, therefore, a warning was unnecessary, he became sufficiently concerned (after precious seconds had gone by discussing the situation with his supervisor) to radio the Cessna pilot that the PSA jet was in his vicinity, descending towards the airport and, reassuringly, had the Cessna in sight.

Too late.

Simultaneously with the last transmission, the descending 727 jetliner, under weather conditions consisting of a clear sky with excellent visibility, crunched the small Cessna!

According to witnesses, both aircraft were proceeding in an easterly direction. Flight 182 was descending and overtaking the Cessna, which was climbing in a wing-level attitude. (The Cessna has a single overhead wing which blocks the view from the above rear, making it impossible for the passenger jet to be visible to the two pilots of the small airplane.) Just before impact, as Flight 182 banked to the right slightly, the Cessna pitched noseup and was shattered by the right wing of the jet airliner at the 2600-foot level.

Stated one of the hundreds of people who looked up on hearing the explosion following the mid-air collision, "The small plane just blew up and the engine on the right side of the PSA jet burst into flames."

Another witness said, "I thought it was a loud sonic boom. But it seemed too loud and too close. I opened the shades and looked up. All I could see was black smoke and pieces of things falling from the sky."

After the collision, Flight 182, with the flight crew frantically trying to maintain control, began a shallow right descending turn, leaving a trail of vaporlike substance from the right wing. A bright orange fire erupted in the vicinity of the right wing and increased in intensity as the aircraft descended. The airplane remained in a right turn and both the bank and nosedown position increased during the descent to about 50 degrees, at which angle it smashed straight into the ground with a devastating impact.

Both the 727 and the Cessna were completely destroyed. There were no survivors.

Both aircraft crashed during daylight hours right smack into a residential community of small, old houses inhabited by mostly elderly, retired persons. The disaster area is located about three miles northeast of Lindbergh Field.

It took 20 seconds for the airliner, after the collision with the Cessna, to impact with the ground. Those 20 seconds must have seemed like eternity to the passengers and crew helplessly trapped in the out-of-control aircraft turning and diving towards the earth.

It even appeared so to a spectator, who said, "I don't know why but, God, it seemed like it took forever to fall."

Another witness stated, "It hit the ground like a big atomic explosion—there was debris falling everywhere. It's tragic," he said. "There are pieces of bodies in the alleys and back yards."

This normally quiet neighborhood was in complete turmoil as parts of the aircraft, flaming debris and bodies rained down on its homes, back yards and roads. Residents standing in the streets were in shock, as were those who ran out of their homes in bewilderment at the frightening sounds of wreckage and human remains impacting the area.

A local radio station was able to dispatch a reporter to the scene within minutes of the crash and place him on the air, live, so that listeners would hear the details of the immensity of the tragedy.

"I can't describe it," said the emotionally upset reporter.

"Pieces of bodies are everywhere. Some on rooftops, some in trees. Homes are on fire. People are running everywhere. This is the worst tragedy I have every seen."

All seven crew members and the 128 passengers from the PSA jet airliner as well as the two pilots of the small airplane were casualties. In addition, seven persons on the ground were killed from the falling debris and subsequent fires.

Some residents of the community were treated for shock in the hospital along with others sustaining physical injuries. Property damage was also heavy; in addition to automobiles that were struck, 22 dwellings were severely damaged or destroyed.

Many of the bodies were difficult to identify because of the severe burns and mutilation from impact. In some cases, dental records had to be used. The county coroner's office asked friends and relatives of suspected victims to "try to be patient and understanding," since it might take several days before positive identifications could be made.

Those who were waiting at the gate to greet the arrival of Flight 182 initially noticed the "delayed" announcement on the visual screens. Shortly afterward, they were told there was an accident and were given emergency numbers to telephone for further information.

More than 40 airline personnel were assigned the unpleasant task of notifying the next of kin. "When they were hired, they knew their duties might some day include this kind of job," said the director of public relations for the airline. "There was no special training for the work. It's hard, gruesome, tough."

As fate would have it, aboard the fatal flight were 11 "space available" passengers who had learned at the last minute that there was a seat available for them. In addition, there were 18 non-paying passengers, most of them airline employees commuting in the course of their work—an additional blow to Pacific Southwest Airlines.

When the news of the crash reached passengers at Lindbergh Field waiting to board the same airplane for the return flight to Los Angeles and San Francisco, some were hesitant about continuing their plans to fly that day. "It seems like it was only ten seconds later and they had another plane ready," remarked a passenger waiting to fly to San Francisco.

She and her friend considered cancelling their trip. However, she added, "She convinced me that there couldn't be two [crashes] in one day."

Others at the airport in San Diego, trying to reassure themselves, echoed the same sentiments. "Statistically, it seems impossible," stated another shaky would-be traveler.

In an attempt to piece together the fateful events that ultimately led to this tragedy, there are a number of observations worth noting:

1. There is a "conflict alert" system at San Diego to alert controllers of a potential collision between two or more aircraft. This consists of both a visual alarm—the letters "CA" (conflict alert) blinking on the controller's display screen—as well as a five second aural alarm sound.

One of the problems of this system is that the San Diego facility experiences an average of 13 conflict alerts per day, many of which are nuisance alerts where there is no actual conflict or no aircraft close enough to require further action. Unfortunately, after a while, these warnings become part of a "cry wolf" or "false alarm" syndrome.

When the approach controller heard and saw the PSA 182-Cessna conflict alert 19 seconds before the actual collision, he discussed the situation with his coordinator. He said he had pointed out the traffic to Flight 182; the flight crew had stated that they had the traffic in sight and would maintain visual separation from the Cessna. Therefore, as far as the controller and his coordinator were concerned, there was no conflict, and no further action was required. However, at about the same moment of the collision, the controller evidently had second thoughts and did advise the Cessna again that Flight 182 was in his vicinity and had him "in sight." Too bad, a precautionary command to Flight 182 to "climb immediately" only ten seconds earlier, might have prevented the disaster.

2. *The Airman's Information Manual* published by the FAA states that when a pilot has been told to follow another aircraft or to provide visual separation from it, he should promptly notify the controller if he does not sight the other aircraft.

The chief pilot of Pacific Southwest Airlines testified that upon receipt of a traffic advisory, the company requires the flight crew to "look for the traffic until you sight him or acknowledge that you do not have him in sight." If the traffic was lost from sight, the controller was to be immediately advised of that fact. Had the PSA crew radioed "We've lost our traffic," or, "Can't locate that Cessna," there would have been plenty of time for a now alerted and alarmed traffic controller to issue avoidance instructions.

3. Sometimes, the testimony of observers cannot be relied upon. Fifteen witnesses said they saw a third aircraft (which could have confused the PSA 182 crew) in the vicinity of the collision. However, two of these witnesses described an aircraft heading north, three described an aircraft heading west, six described an aircraft heading east, and four saw an aircraft but were unable to place it on any heading.

4. A visibility study showed that a view of the Cessna from the PSA jet would have been positioned at the bottom of the pilots' windshields just above the wiper blades. Depending upon the eye reference points, the Cessna could have been masked by the 727's cockpit structure. Therefore, it is possible they could not see it unless they either leaned forward or raised their seats, or both.

As for the Cessna, because of its high wing, the two pilots could not possibly see the three-engine jet descending into them from the rear. Also, since the Cessna pilots were told that they were being overtaken by an aircraft whose flight crew had them in sight, it would be unrealistic to conclude that they would have made any attempt to turn their aircraft in order to sight Flight 182. As a matter of fact, a specific flight rule states, "Each aircraft that is being overtaken has the right of way and each pilot of an overtaking aircraft shall alter course to the right to pass well clear."

5. An unusual aspect of this accident is that both aircraft were operating under radar surveillance and were observed up until the point of collision. Evidence indicates that flight crews in a "see and avoid" environment, usually exercise a lower vigilance in areas where they receive radar assistance than in non-radar areas. They seem to rely on the radar controller to point out aircraft that

may be in conflict with theirs, instead of being vigilant on their own in seeking out and maintaining visual contact with other aircraft.

6. Newspaper reports initially placed the blame for the accident on the Cessna. Stories were printed (with diagrams) explaining how the student pilot on the small Cessna aircraft, practicing flying blind under a hood, rammed the unsuspecting 727 jet from behind causing this terrible collision. It took a while to sort out the facts and come up with an accurate assessment of this accident.

As happens in all major aircraft crashes, the National Transportation Safety Board was notified of the accident and immediately dispatched an investigative team to the scene. After a thorough investigation, the Safety Board issued the following release as to the cause of the accident:

"The . . . probable cause of the accident was the failure of the flight crew of Flight 182 to comply with the provisions of a maintain-visual-separation clearance, including the requirement to inform the controller when they no longer had the other aircraft in sight."

The Safety Board also stated that contributing to the accident were the air traffic control procedures which allowed the controllers to instruct both airplanes to fly by the visual (see and avoid) flight rules, when they had the capability, via their radar equipment, to safely keep both aircraft apart.

The report on this accident would not be complete without reading the transcript of the last few minutes of conversation of the crew with each other and the controllers. This was recorded on tapes of the Cockpit Voice Recorder recovered from the wreckage of the PSA 727 jet. For the sake of brevity, some transmissions were omitted. In addition, some non-pertinent words and expletives were deliberately deleted.

### TIME 8:59:11 (seconds) AM

Off-Duty PSA Captain: Are we there yet?
Co-Pilot:                    Just about.
PSA Engineer:              Landing, turn off lights.

| | |
|---|---|
| Captain: | On. |
| Engineer: | Seatbelt sign on too. |
| Co-Pilot: | It's on. |
| Engineer: | Fuel, shoulder harnesses. |
| Captain: | On. |

TIME 8:59:30 AM

| | |
|---|---|
| San Diego Approach Control: | PSA 182, traffic 12 o'clock [straight ahead], one mile northbound. |
| Captain: | We're looking. |
| San Diego Approach Control: | PSA 182, additional traffic's 12 o'clock, three miles just north of the field northeastbound, a Cessna one seventy-two climbing out of one thousand four hundred. |
| Co-Pilot: | Okay, we've got that other 12. |
| San Diego Approach Control: | PSA 182, traffic's at 12 o'clock, three miles, out of one thousand seven hundred. |
| Co-Pilot: | Got 'em. |
| Captain: | Traffic in sight. |
| San Diego Approach Control: | O.K., sir, maintain visual separation, contact Lindbergh tower . . . have a nice day now. |
| Captain: | Okay. |

TIME 9:00:34

| | |
|---|---|
| Captain: | (to tower) 182 [is] downwind [on approach to runway]. |
| Tower: | PSA 182, [this is] Lindbergh tower, traffic 12 o'clock, one mile, a Cessna. |
| Captain: | (cockpit conversation) Is that the one [we're] looking at? |
| Co-Pilot: | Yeah, but I don't see him now. |
| Captain: | (to tower) Okay, we had it there a minute ago. |
| Tower: | 182, roger. |
| Captain: | I think he's passed off to our right. |

| | |
|---|---|
| Tower: | Yeah. |
| Captain: | (cockpit conversation) He was right over there a minute ago. |
| Co-Pilot: | Yeah. |

### TIME 9:01:10

| | |
|---|---|
| Tower: | PSA 182, cleared to land. |
| Captain: | 182's cleared to land. |
| Co-Pilot: | (intra-cockpit) Are we clear of that Cessna? |
| Engineer: | Supposed to be. |
| Captain: | I guess. |
| Unidentified voice: | (sound of laughter) |
| Off-Duty PSA Captain: | I hope. |
| Captain: | Oh yeah, before we turned downwind, I saw him . . . probably behind us now. |
| Co-Pilot: | Gear down. |
| | (sound of clicks and sound similar to gear extension) |
| | (sound of nose gear door closing) |

### TIME 9:01:45

| | |
|---|---|
| Captain: | Whoop! |
| Co-Pilot: | Aghhh! |
| | (sound of impact) |
| Off-Duty PSA Captain: | Oh #### [deleted word] |
| Unidentified voice: | #### [deleted word] |
| Captain: | Easy baby, easy baby. |
| Unidentified voice: | Yeah. |
| Captain: | What have we got here? |
| Co-Pilot: | It's bad. |
| Captain: | Huh? |
| Co-Pilot: | We're hit, man, we are hit! |
| Captain: | Tower, we're going down, this is PSA. |
| Tower: | Okay, we'll call the [emergency] equipment for you. |

TIME 9:01:58

| | |
|---|---|
| Unidentified voice: | Whoo! |
| | (sound of stall warning) |
| Unidentified voice: | Bob [co-pilot's name]. |
| Captain: | [to tower] This is it, baby. |
| Co-Pilot: | #### [deleted word] |
| Unidentified voice: | #### [deleted word] |
| Captain: | Brace yourself. |
| Unidentified voice: | Hey baby. |
| Unidentified voice: | Ma, I love ya. |

That last poignant cry was recorded on the Cockpit Voice Recorder exactly two and a half seconds before the airline jet crashed nose down into the ground.

To a Mother? Wife?

We'll never know.

# Chapter 4

# Landed on the Wrong Runway

The accident to PSA 182 would not have occurred if at least one of the crew members in the cockpit of the airliner had paid more attention to the warnings of the traffic controller. Coincidentally, we have another situation in which the repeated instructions of an air controller appear to have been ignored.

If you were pilot of a commercial airliner, how many times would you have to be told that a runway is closed for landing before you believe it? (Especially when you have, besides a concern for your own personal safety, the responsibility for the lives of 88 others.) Would once be enough? It certainly ought to be. If not, how about twice? That should absolutely get your attention. No? Well, try three times, and—you won't believe this—that didn't do the trick. That pilot was determined to land on the wrong runway.

How could that happen? We'll show you. Just follow the flight of another DC-10, with 89 persons aboard, that took off from the Los Angeles Airport at the unearthly hour of 1:40 in the morning on October 31, 1979. The destination of the Western Airlines aircraft was Mexico City's 7,000 foot-high Benito Juarez Airport. This flight, number 2605, regularly scheduled for departure after midnight and arrival at Mexico City before dawn, was known as El Tecolate, or the Night Owl.

The Captain was 53 years old and had logged a considerable number of flying hours on DC-10 aircraft. Besides his long years of experience, he had accumulated 28 landings as pilot-in-command into Mexico City. In addition, both his First Officer and Second Officer were also credited with many landings into Benito Juarez Airport. Obviously, they knew that airport fairly well (although you wouldn't think so based upon what happened later).

The flight was routine and uneventful until the DC-10 approached Mexico City. At this point, Flight 2605 received radio instructions from a controller, which included the information that the runway it was to land on was 23R *(Right)*.

Benito Juarez airport has two parallel runways, 23L (Left) and 23R (Right) which are separated from each other by 200 yards. However, that day, 23L was not in use. Twelve days prior to this flight, a Notam (Notice to Airmen) was issued to inform airlines who used the International Airport of the City of Mexico that runway 23L was closed for repairs. Western Airlines, in turn, advised all of its crews flying to Mexico of the contents of that Notam. In addition, the airline gave the crews a specific flight plan which carried its own notice that runway 23L was closed because of construction work.

After the controller instructed Flight 2605 to land on runway 23R, he followed with this radio message, "Western 2605, tower advised ground fog on the runway and two miles visibility on the final approach."

The report of ground fog and poor visibility did not disturb the flight crew. They were used to it at this airport, which is located at a high altitude between mountains. Nevertheless, despite their confidence, the Mexico City Airport has been criticized because it lacks a Runway Visual Range System—a device that assists an approaching pilot in determining the visibility at the beginning of the runway.

The lack of this system is one reason why the Mexico City

Airport has been given a "red star" rating by the International Federation of Air Line Pilots Association, which, among other functions, rates the safety of airports around the world. This rating means that the pilots have judged this airport to be seriously deficient in some areas.

Getting back to Flight 2605 as it is approaching the airport, we listen to the following transmissions taking place between the DC-10 and the Mexico City tower (with explanations in brackets for clarification):

| | |
|---|---|
| Aircraft: | Good morning, Mexico tower, Western 2605 is inbound for 23 [not specifying Left or Right runway]. |
| Tower: | Western 2605, 23 *Right* [specifying the correct runway—now the second time the flight was so informed], report over Mike Echo [location point on map] wind calm. |
| Aircraft: | Western 2605 is inside Mike Echo. |
| Tower: | Western 2605, advise [when] runway in sight [so we can clear you to land]. |
| Tower: | (Cautioning) Western 2605, you are to the left of the track. |
| Aircraft: | (Admitting) Just a little bit. |
| Tower: | Advise [when] runway in sight . . . there is a . . . layer of . . . fog over the field. |
| Aircraft: | (Acknowledges) 2605, Roger. |

(Meterological information taken at the Mexico City Airport for that time period indicated that the sky was obscured and visibility was considerably reduced by smoke, haze and fog.)

| | |
|---|---|
| Tower: | 2605, do you have approach lights on [23] Left in sight? |
| Aircraft: | Negative. |
| Tower: | Okay, sir, approach lights are on 23 Left—*but that runway is closed to traffic!* |
| Aircraft: | (Acknowledges) Okay, 2605. |

At this point, Flight 2605 has been reminded, via three separate transmissions, that runway 23L was closed and the runway to land on was 23R. The experienced flight crew acknowledged this information from the controllers and now, with the radio silent, concentrated on making its instrument landing.

Afterwards, according to the information obtained from the voice recorder, it confirmed that the crew was performing an instrument approach procedure using runway 23L with a "sidestep" to line up and land on 23R. (This is a commonplace instrument maneuver used at many airports with parallel runways.) However, regardless of whether 23L or 23R is used, a "missed approach" and a "go-around" is mandatory at this airport if the runway is not visible when the aircraft gets down to 600 feet.

What actually happened in this instance is that the DC-10 continued the approach without the required procedure of a crew member, who was not flying the airplane, calling out the descending altitude and announcing when the minimum decision height of 600 feet has been reached. Despite instructions from the tower to advise it when it had the runway in sight, Flight 2605, in silence, continued its descent through the critical 600 foot altitude (with no one knowing whether or not the runway was visible to the crew at that level) right down to the moment of touchdown—on the wrong runway!

At 5:41 AM, Flight 2605 touched down slightly to the left of the closed runway, 23L. Its left gear alighted on the grass and its rear gear settled on the shoulder, both left of the runway. Approximately 300 feet after touchdown, veering slightly right, the DC-10 finally entered the runway. An FAA official stated afterwards, "The important thing is that he should never have come down until he saw the runway light. . . ." The official said, "The runway was covered with fog but it looks like he [the pilot] just tried to dive for it."

The DC-10 was hurtling down the wrong runway when the flight crew noticed, in a momentary break in the heavy fog, that a repair truck loaded with 10 tons of soil was moving on the left shoulder of the closed runway. The crew applied maximum takeoff power and tried to lift the nose of the aircraft in a desperate attempt to clear the heavy construction vehicle. The jumbo jet couldn't quite make it.

Although Flight 2605 became airborne again, its right main landing gear smashed into the left side of the truck. At impact, the DC-10's right main landing gear became detached, blowing three of its four tires, fragmenting the rims and breaking the axle and front brake assemblies. In addition, the right lower sections of the tail, as well as the right wing flaps, were torn off.

The truck itself was completely demolished. The wreckage of that vehicle was fragmented and scattered over a large area.

Even after the violent collision with the truck, the DC-10 remained airborne. Still about six feet off the ground, the airplane rolled to the right where it now struck and destroyed the cab of a mechanical excavator located on the right side of the runway.

The jumbo jet then drifted to the right of runway 23L and increased the roll until the right wing started to drag on the ground leaving deep scratch marks and destroying ground lights and telephone junction boxes. At this point, fire started to break out in the airplane.

As the DC-10 continued its half-flying, half-ground-scraping journey rolling to the right, it actually reached and crossed its properly designated runway, 23 Right. Still moving in its right-tilted position, the right wing struck a corner of a repair-hangar where the fuel tanks ruptured.

The badly-wounded DC-10, now spraying burning fuel, continued its agonizing journey until the aircraft ploughed head-on into a concrete-block storage building housing spare parts, vehicles and office equipment for Eastern and Pan American Airlines. This was the main impact; the building, engulfed in flames, was completely demolished. In addition, the momentum of the aircraft was such that part of it came to rest outside the airport boundaries, some on top of a nearby house which also caught fire.

Immediately after the fiery airplane collided with the concrete-block building, the following transmissions took place:

| | |
|---|---|
| Tower: | Tower sub-station, Western just crashed. |
| Sub-station: | We saw it, we are in Room 6. |
| Tower: | Please advise the fire department and the command. |

Sub-station: Okay.

The wreckage came to rest about 400 feet from the airport's fire station. Fire personnel responded immediately and were later joined by the firemen from the Mexico City central fire station.

When parts of the flaming debris of the aircraft fell on the houses located near the airport boundary, terrified residents leapt from their beds and ran out into the fog. "We were all asleep," said one of them. "We heard a terrible noise and an explosion. People were shouting, 'Run for your lives!' "

On the aircraft itself, strangely enough, the tail section did not burn. One of the stewards, an amateur weightlifter, was seated in the rear of the aircraft and was able to assist survivors who were also seated in that section. It was reported that, using only his bare hands, he widened an opening in the aircraft fuselage sufficiently so that passengers could squeeze out.

One passenger, a pilot for a private aviation company, remembered that everything seemed quite normal—until the landing.

"I felt the wheels had touched but they touched very hard," he said. "A few of the oxygen masks fell out, but I thought we had already landed O.K. Then the right wing touched the ground and it started to disintegrate.

"It just fell apart," he recalled. Fortunately, he was seated in the only section of the fuselage which did not splinter into fragments.

He crawled toward a gaping, blazing hole nearby, being burned on his hands, his face and his hair on the way. He fought his way outside and staggered towards ambulances whose sirens he could already hear.

A day later, from a hospital bed with his hands heavily bandaged, he expressed his thanks for having survived and voiced one regret. "My mustache," he mourned. "My Pancho Villa mustache. All burned off."

Another survivor, who had planned to spend a month's vacation in Mexico, took stock of his injuries and says it could have been worse.

"I was on the right side, over the wing. I sat there [in the smoking section] because I smoke, and that's what saved me. It's terrible that something like that can save your life."

Of the 13 crew members serving on board El Tecolate, all three flight crew members and eight out of the 10 flight attendants were fatality injured. Two flight attendants escaped with minor injuries.

Of the 76 passengers making the flight, 61 perished, 13 suffered serious injuries and two, miraculously, escaped without injury. In addition to these numbers, we have to add one more fatality, that of the truck driver.

Back in Los Angeles, relatives of passengers and crew members of Flight 2605 awoke to the news that morning that the DC-10 had crashed at the Mexico City Airport.

Those unable to get through by phone to Western Airlines rushed to the airport seeking news. However, when they arrived, around 7:45 AM, no one there could answer their questions. They were told to call an extension on a white telephone nearby on the ticket counter.

But that line was busy, too. The distraught relatives, their faces mirroring their anxiety, dialed and re-dialed the number in their quest for information.

One man feared his wife and daughter were aboard Flight 2605. One woman, who believed her husband was on that flight, was practically unable to speak, her face frozen in grief and fear.

Another woman did not know the number of the flight her son, age 27, had taken to Mexico City. He had earned a master's degree at California State University and had been studying languages trying to perfect his Spanish to help him in his work as a counselor.

"We heard about the crash on the radio," she said as she continued to dial the white telephone. "We couldn't get through on the phone so we came on down. We're frantic."

When she finally did get through, she learned that her son had been a passenger on that flight. Now began the agonizing wait to find out if he was one of the 17 survivors.

Since there was no special place to wait, she went upstairs to the cafeteria. "The waiting is the worst," she said as she drank a cup of coffee.

By 9:00 AM, Western had made arrangements to gently escort the relatives to the airline's plus Horizon Club, where food and coffee were brought to them. At least, it removed them from an almost ghoulish atmosphere in the terminal, where some people were outfitted in Halloween costumes.

Western Airlines also set up a communications headquarters across the road from the terminal. Eight airline staff members sat at a bank of telephones answering hundreds of calls, many from people who were not sure a relative or friend had taken Flight 2605 that morning.

Occasionally, good news would be forthcoming. "I'm happy to inform you he was not aboard the flight," one of the employees told a caller.

As the names of the victims became known, airline officials began the unhappy task of notifying the next of kin. One woman, taking courses at a university, was called out of class and told that her husband had been killed in the crash. She went home to pray and comfort her five children. Then, several hours later, the airline telephoned her and told her that her husband, seated in the tail end of the aircraft, although severely injured, had survived.

"He's alive because he always sits in the back of the plane and because he's always late for everything, so he *has* to [sit there]," she said. "Thank God."

Not so fortunate was the woman who was waiting for word of her counselor son. The news finally came. He was among the dead.

Another fatally-injured passenger was a highly-experienced reporter on his way to cover rioting in El Salvador. He had previously supervised ABC's coverage of the Nicaraguan revolution and served as a field producer for the network's coverage of the American Airlines DC-10 crash in Chicago, the worst single airplane disaster in U.S. aviation history.

The irony of the situation is that, after covering the Chicago DC-10 crash, the reporter decided that from now on he would have nothing to do with this jumbo commercial airliner. His bureau chief recalled that, only recently, his ace reporter told him, "I have not flown on a DC-10 since the crash in Chicago and I won't fly a DC-10 unless it's the only way I can get there."

This reporter, a Latin American expert, originally was not even supposed to go to El Salvador. ABC News pulled him off another assignment to cover this important story.

Evidently, flying on the DC-10 was the only way he could get there.

Western Airlines offered to fly the relatives of the passengers of Flight 2605 to Mexico City. Obviously, there was a contrast of emotions among the next of kin on these flights.

Those whose relatives perished, appeared at the Mexican government's morgue to complete the task of identifying the 72 victims. The grieving families took turns viewing the bodies lying on plastic-sheeted floors at the Medical Forensic Hospital and claimed their dead.

At the same time, on the other hand, other families were arriving at the hospitals to give thanks. They were meeting with the living—their relatives who had, somehow, survived.

The investigation of an accident is generally made by the government of the country where the accident occurred. However, the agreements of the International Civil Aviation Organization provides for participation by the government of registry of the aircraft. Therefore, the U.S. National Transportation Safety Board participated in the Mexican Government's investigation of the Western Airlines DC-10 accident in Mexico City.

After thoroughly investigating the events leading up to the crash, they came to these conclusions:

1. Western Airlines had advised all of its crews concerning the closing of runway 23 Left because of construction work.

2. After contacting the Mexico tower, the crew was issued pertinent instruction and information for landing at runway 23 Right—which the crew acknowledged.

3. The crew descended below the 600 foot minimum altitude without reporting the runway in sight and failed to initiate the missed approach procedure.

4. The crew never advised the tower they had the runway in sight; consequently, they were not cleared to land.

5. According to the cockpit voice recorder readout, the crew did not follow the procedures concerning altitude call-outs during the final approach.

In view of the above, the investigative authorities stated the probable cause of the accident to be the following:

"Failure of the crew to adhere to the minimum altitude for the approach procedure they were cleared for. Failure of the crew to follow approved procedures during an instrument approach. Landing on a runway closed to traffic."

One last comment. As disastrous as this accident was with its relatively high total of fatalities, it could have been far worse.

Four commercial jet airliners were sitting at the Mexico City Airport terminal building ready to take on passengers. Airport officials stated that, if the DC-10 had not clipped its wing on an airport hangar and spun away, its original trajectory would have taken the out-of-control flaming aircraft right into the four fully-fueled jets parked at the terminal. And that would be something to write about.

Chapter 5

# Ran Out of Fuel

We have another almost unbelievable accident to write about. This, too, occurred in the landing process and also appeared to be as unlikely an accident as could happen. But, happen it did.

On December 28, 1978, a United Airlines four-engine jet on a flight to Portland, Oregon, ran out of fuel and crashed.

The DC-8 aircraft, carrying 189 persons, was only six miles away from the destination airport when the accident occurred. The irony of the situation is that the flight had circled within a short distance of the airfield for approximately one hour, and deliberately delayed its arrival to try to cope with a landing gear problem. A mere five minutes less time spent in the holding pattern would have saved enough fuel to assure the airliner of safely reaching the runway.

Let us follow the flight path of that heavy airplane to see what transpired that day to cause such an unlikely disaster

The airliner, known as Flight 173, departed Denver, Colorado, at 2:47 PM. The United Airlines aircraft carried eight crew members who were responsible for the comfort and safety of 181 passengers, including six infants. The planned time for the trip from Denver to Portland, Oregon, was less than two and a half hours.

Flight 173's tanks contained sufficient fuel for the flight to Portland, plus extra fuel to meet FAA and company requirements. This reserve fuel amounted to 60 additional minutes of flying time to contend with unforeseen delays.

Available to the flight crew was a monitoring system to check the fuel consumed along the intended flight route. This consisted of a computer printout which predicted the amount of fuel that would be used between several points en route. The crew would be able to check the actual fuel used against the predicted fuel use at each of these points.

The flight to Portland was smooth and routine. Shortly after 5:05 PM, right on schedule, Flight 173 called the Portland Approach controller and reported that it had the field in sight. Portland instructed the flight to descend to 8,000 feet and, shortly after, to continue its descent.

During a post-accident interview, the Captain stated that, as the landing gear extended, ". . . it was noticeably unusual and . . . it seemed to go down more rapidly . . . it was a thump, thump in sound and feel . . . much out of the ordinary for the airplane. It was noticeably different and we got the nose gear light but no other light."

Flight attendant and passenger statements also indicated that there was a loud noise and a severe shudder.

"We were making a normal approach and I remember the pilot announcing we were at 5,000 feet just before the gear went down," said one of the passengers. "The gear made a terrible, terrible jolt. We knew immediately that something was wrong. The pilot came on the intercom and said he was concerned and that we'd circle around while the crew did some checking. . . ."

The Portland Approach controller, unaware of the fact that Flight 173 was experiencing a gear problem, radioed the jet airliner to now contact the tower for landing instructions. The aircraft responded, ". . . negative, we'll stay with you [Portland Approach]. We got a gear problem. We'll let you know." This was the first indication to anyone on the ground that Flight 173 had a problem. Portland Approach acknowledged and told the

United pilot to maintain five thousand feet.

At 5:15, two minutes after the scheduled arrival time, Portland Approach advised, "United one seventy-three heavy, turn left heading one zero zero and I'll just orbit you out there 'til you get your problem." Flight 173 acknowledged the instructions without any feeling of urgency. After all, the aircraft's tanks did contain another 60 minutes of fuel.

However, lets see how the flight crew proceeded to use up this one hour reserve while Portland Approach was vectoring the aircraft in a holding pattern within a 20-mile radius of the airport.

For the first half of the 60 minutes, the following took place:

The flight crew discussed and employed emergency actions available to them to try to get the landing gear locked in the full down position (not knowing whether it would collapse on touch-down).

The chief flight attendant entered the cockpit to ask for instructions and, after some discussion, was assured by the Captain that he would let her know what he intended to do.

The Captain radioed the United Airlines Maintenance Center in San Francisco and explained the problem to company maintenance personnel. He advised that they had about 7,000 pounds of fuel remaining (of the original 46,700 pounds) and stated his intention to circle in the holding pattern for another 15 or 20 minutes.

All of this activity used up 30 minutes of the reserve fuel. Now, at 5:44 PM, United Maintenance asked, ". . . United 173, you estimate that you'll make a landing about five minutes past the hour [of 6 PM]. Is the okay?" The captain responded, "Ya, that's good ball park. I'm not going to to hurry the girls. We got about a hundred sixty-five people on board and we . . . want to . . . take our time and get everybody ready and then we'll go. It's clear as a bell and no problem." (Note that the captain set a time of 6:05 PM to land the airliner.)

We are now into the last half-hour of fuel. Instead of a sense of urgency, the conversations in the cockpit seem to be getting longer. At this stage, we will employ a countdown to follow what is taking place while the dwindling fuel is being guzzled up by the four thirsty jet engines.

Thirty minutes to empty tank—The Captain and the chief flight attendant talked about the time needed to prepare the passengers in the cabin should an emergency evacuation become necessary.

Meanwhile, the First Officer and flight engineer noticed that the aircraft was down to 5,000 pounds of fuel. This information was passed on to the Captain, who acknowledged it, but did not appear to be concerned.

Twenty-five minutes left—The Captain told the flight engineer to figure on another 15 minutes of circling to which the engineer protested, "Not enough. Fifteen minutes is gonna—really run us low on fuel here." The Captain remained unperturbed. (It is ironic that the only cockpit occupant who perished in the later crash, was the one who actually voiced his apprehension about the fuel situation—the flight engineer.)

Approximately 20 minutes left—The Captain instructed the flight engineer to inform the company representative at Portland Airport that they would land with about 4,000 pounds of fuel. The engineer complied and then turned to query the Captain, "He wants to know if we'll be landing about five after." The Captain replied, "Yes." (This is the second time that the Captain stated he would be touching down on the runway at 6:05 PM).

The fuel gauges now indicated that only 4,000 pounds of fuel remained, 1,000 in each tank. The Captain now decided to send the flight engineer to the cabin to ". . . kinda see how things are going . . ." Meanwhile, the airplane was droning on at 5,000 feet, using up more of the precious fuel.

Approximately 15 minutes left—The engineer returned to the cockpit and reported that the cabin would be ready shortly. Looking at the gauges, the flight engineer called out, "We got about three [thousand pounds] on the fuel and that's it!" The Captain acknowledged. The airliner was now at the closest point to the airport, *only 5 miles away.*

The Portland Approach controller radioed and asked for a status report. The First Officer replied that they intended to land in about five minutes and would like the airport emergency equipment to be standing by. The aircraft was now moving *away* from the airport.

Approximately ten minutes to empty—For several minutes, up

to 6:06 PM (past the time the Captain had twice stated he would touch down), the flight crew started another discussion on cockpit procedures to be followed before and after landing.

Finally, the Captain said, "Okay. We're going to go in now. We should be landing in about five minutes." And then it happened. Almost simultaneous with the Captain's decision, the First Officer said to the flight engineer, "I think you just lost number four [engine]."

It was now almost 6:07 PM. The First Officer turned to the Captain, "We're going to lose an engine. . . ." The captain, puzzled, replied, "Why?" The First Officer again stated, "We're losing an engine." The Captain, evidently, could not believe what he was hearing. Again he asked, "Why?" His first officer's comment was succinct. "Fuel."

The captain immediately called the controller and requested a clearance for an approach to the runway. This was the first request for an approach clearance from Flight 173 since the landing gear problem began. As luck would have it, the aircraft was now *19 miles away* from the airport, almost at the extreme end of the circling path.

With one engine flamed out because of fuel starvation and with the realization that the other engines couldn't be far behind, the crew's cool, relaxed attitude started to show some signs of coming apart. From 6:07:27 until 6:09:16, the cockpit voice recorder indicated that the conversations in the cockpit varied from exhortations from the Captain to ". . . get some fuel in there . . . you gotta keep 'em running . . ." to the First Officer pleading, ". . . get this * [deleted] on the ground" followed by the flight engineer defensively explaining that, despite his efforts, the tanks were almost ". . . empty."

At 6:12:42, the captain, nursing the airliner towards the airport, asked the Portland Approach controller how much farther he had to go. Portland Approach responded, "Twelve flying miles," and told the flight to contact Portland tower for landing instructions.

At 6:13:21 (more than eight minutes after the originally announced touchdown time), the flight engineer reported that, now, two engines had flamed out. Both the Captain and the First

Officer finally realized that it was hopeless—they could never make the airport before the remaining two engines would be lost.

The Captain told the First Officer. "Okay. Declare a Mayday [emergency]." The first officer radioed, "Portland tower, United one seventy-three heavy, Mayday. We're—the engines are flaming out. We're going down. We're not going to be able to make the airport."

At about 6:15 PM, only six tantalizing miles short of the runway, Flight 173, with all of its fuel exhausted and no thrust available, descended into a forest in the midst of a populated area of suburban Portland.

The aircraft first struck two trees about 100 feet above the terrain, and then, in its downward path, continued to strike tree after tree, closer and closer to the ground. Barely missing a 47-unit apartment complex, the aircraft smashed and destroyed an unoccupied house before hurtling into a five-foot embankment next to a city street. Continuing its relentless journey across the street, it finally came to rest between several trees and smack on top of another unoccupied house.

The aircraft was totally destroyed by the crash; there was no fire. The two unoccupied homes were also destroyed, as well as telephone poles and high-tension powerlines. Although there were many occupied houses and apartment complexes in the immediate vicinity of the accident, there were no ground casualties and no post-crash fire.

Considering the fact that this crash slammed into a grove of trees, the casualty rate was extraordinarily low—out of a total of 189 persons on board, only ten deaths. No doubt, a contributing factor was that all the fuel had been consumed, thus preventing a fire with its probable resulting explosion. In addition, 23 occupants were seriously injured, with 156 escaping with minor injuries.

The 10 occupants killed in the crash were located between the flight engineer's station in the cockpit and row 5, all on the right side of the cabin. That section of the aircraft was destroyed during the sequence of impacts occurring while crashing.

Newspaper reports stated that the Captain, who had 27 years of experience flying with United, felt all along he had sufficient fuel to reach the airport right up until the engines flamed out. He said that while trying to determine the extent of the gear problem, as a safety measure he was attempting to use up excess fuel to minimize the danger of a fire that might erupt from a possible crash landing.

With all four of his jet engines unexpectedly flaming out, the Captain was faced with making a quick decision. As the extra-heavy passenger aircraft without power became a reluctant glider and started to settle rapidly, he had only 75 seconds left before the huge aircraft would run out of altitude and touch down somewhere below in the darkness of night.

The pilot did have one thing going for him—he had flown this route many times and knew the area well. He discarded the idea of landing on an interstate highway because of the possibility of rush-hour traffic. He also decided against ditching into the nearby Columbia River because of its swift current and icy waters.

"There were several possibilities, none of them good," he said. "I found this one spot—a grove of trees with houses on each side." Because that area was dark and there were no lights, he assumed there were no houses among the trees. He said he took the plane in, nosing it between two trees to cushion the impact. Afterwards, he remarked, "I never went to school on how you land in trees. . . ."

"It's just amazing," said one of the policemen on the scene. "If they had to crash, they couldn't put it down in a better place."

As for the passengers in the aircraft, the Captain gave much of the credit for the high survival rate to the chief flight attendant who was killed in the crash. "She was in charge in preparing the cabin and she did it in such a good way," he said. "She was a damn good stewardess."

In addition to the death of the chief flight attendant, the flight engineer was also killed in the crash. A hospital spokesman, at the time, stated that the Captain was recovering from a broken ankle, broken ribs, serious head cuts, and back injuries, while the First Officer remained in critical condition.

The passengers, although aware of the problem with the landing gear, thought the airplane was landing on the runway. Since they received no warning from the flight crew that the plane was about to crash, they were caught completely by surprise when the aircraft started to contact trees on its descent into the ground.

Said one survivor, "When I felt the first bump, I thought, 'Oh good, nothing went wrong.' Then, all of a sudden—Jesus!"

"We hit hard," said a 19 year-old girl. "We knew we weren't on the runway. At that point a stewardess yelled, 'Grab your ankles.'"

One woman explained why her husband, a Continental Can executive, changed seats with her as the plane circled. "I was next to a middle exit. He thought a man should be there to open the emergency door. . . ."

After the crash, the executive opened the emergency door next to him and started helping people out of the wreckage. He was unaware at the time that his leg was injured.

His wife later cradled one of her husband's shoes. "They had to cut the laces to get it off. His leg was broken," she explained.

Two different families, each consisting of five persons, were dealt completely opposite hands by "lady luck."

One little girl, age three, was the sole survivor in her family of five. Both her parents and two younger sisters, ages two and one, perished in the crash. She, herself, suffered a fractured right leg and head injuries.

On the other hand, all the members of another family of five, survived. One son sustained a broken nose while the mother only received minor cuts on her legs as she scrambled barefoot from the plane. The rest of the family, the father, another son, and a daughter, emerged unscathed.

One can imagine the emotional reunion of this family afterwards, rejoicing in their survival.

Also aboard Flight 173 was an officer of the Oregon Corrections Division escorting a 27-year-old prisoner. This young criminal had served approximately 18 months of his four-year sentence for robbery when he escaped. He had been apprehended by Colorado police and was being returned to the Oregon penitentiary.

Said the officer, "After the plane crash-landed, I jumped down to the ground from the rear exit and he (the prisoner) stayed aboard and handed the people down to me. He was comforting them and assuring them I could catch them.

"When all the people he could find were out, he helped me back into the plane and we looked for more," continued the police captain. "Then a stewardess told us to leave the plane because of the possibility of fire. I heard him call from behind that he was right on my tail. The last time I saw him was when we started for the plane exit. He must have lost my tail."

The Oregon Prison superintendent stated that he hoped the prisoner would turn himself in because of the possibility that parole board officials would take into account his heroism at the crash site. Meantime, although he is not considered to be a dangerous escapee, an all-points police search has been issued for him.

While going through the wreckage, rescuers were amazed to discover something alive moving around in the interior of the airplane. Coming closer, they saw a dog—a full-grown German Shepherd—walking around and exploring the baggage compartment. Although the dog's cage had been demolished, this four-year-old animal was not hurt at all.

His owners were less fortunate in that they were both injured. Although the couple vowed never to fly again, their dog, appearing calm after his rather exciting first flying experience, seemed ready to try again.

At the time the passengers realized the aircraft was about to crash, some of them instinctively turned to prayer. One survivor stated that he leaned forward onto his knees and "prayed with the two ladies next to me." He recalled, "I remember praying that we didn't catch fire . . . explode."

Another survivor who, together with his family, escaped with minor injuries, later told a reporter, "I want you to print this. The Lord Jesus saved our lives. We were praying and the Lord told me that he was going to do a miracle."

A young lady on board had "a premonition something would happen. When the pilot said there was a problem," she said, "I knew we were going to crash. I'm a Buddhist, so—I started saying mantras."

There was another prayer from the ground; this one of thanks, by a family who could have been occupying one of the two demolished vacant houses at the time of the crash.

Because of a dispute with the owner over the electrical wiring, this family moved out of their home only ten days before the accident. Said the lady of the house, who formerly lived there with her husband and two small daughters, "God was up there protecting us when he fixed it so we had to move out of there. We'd all have been killed. We would have been home eating dinner, when the plane crashed."

Residents of the neighborhood opened their homes to the injured. One family took in nearly 30 passengers and tended to their needs. Injuries ranged from a young girl with a bloody face to a seriously injured man who had to be put on the couch. The only damage sustained by the benefactors was bloodstains on their gold carpet.

One other couple said that they felt something shake the house but didn't realize what it was. Said the husband, "We saw all these people walking in the road. I started screaming and telling them to stay away from the power lines. I thought a car had hit a pole.

"They yelled and told me they were from the plane. Plane? I didn't know what they were talking about. Then I realized it was a crash."

His wife added, "They were all thirsty . . . I don't know how many glasses of water I handed out. We gave away blankets. Two sleeping bags—I doubt I'll see them again but that's okay. They were cold. A lot had lost their shoes. This stewardess, she was black, said her feet were freezing. She was barefoot. My husband gave her a pair of his socks."

The National Transportation Safety Board was notified of the accident about 9:30 PM and sent experts immediately to the scene. After conducting a complete investigation, the Safety Board came to the following conclusions:

1. Fuel exhaustion was predictable. The crew failed to equate the fuel remaining with time and distance from the airport.

2. The fuel indicating system accurately indicated fuel quantity to the crew.

3. The fuel gauges were readily visible to the Captain and the flight engineer.

4. The Captain failed to make decisive timely decisions. He failed to relate time, distance from the airport, and the aircraft's fuel state as his attention was directed completely toward the diagnosis of the gear problem and preparation of the passengers for an emergency landing.

5. Neither the First Officer nor the flight engineer conveyed any concern about fuel exhaustion to the Captain until the accident was inevitable.

In addition to these conclusions, the Safety Board stated that it believed that this accident exemplified a recurring problem—a breakdown in cockpit teamwork during a situation involving malfunctions of aircraft systems in flight. To combat this problem, responsibilities must be divided among members of the flight crew while a malfunction is being resolved. In this case, apparently no one was specifically delegated the responsibility of monitoring the amount of remaining fuel.

In order to get the flavor of the preoccupation of the flight crew with the landing gear problem, one has to listen to the tape of the conversations taking place, on the one hand, between the crew members among themselves and, on the other hand, with the air traffic controllers on the ground. The following are excerpts from the transcript of the Cockpit Voice Recorder recovered from the wreckage with some explanations added (in brackets) for the sake of clarity:

### TIME 5:44 PM Pacific Standard Time

| INTRA-COCKPIT | AIR-GROUND COMMUNICATIONS |
|---|---|
| Capt.: [to flight attendant] How you doing? | |
| Flt Att.: We're ready for your announcement | |
| Capt. [do] you have the signal for not evacuate, also the signal for protective position? | |

Flt Att.:   That's the only
            things I need from you
            right now.

Capt.   Okay, what would you
        do? Have you got any
        suggestions about when
        to brace? Want to do it
        on the PA [public
        address system]?

Flt Att.:   I . . . I'll be honest
with you. I've never had one of
these before . . . my first, you
know.

Capt.   All right, what we'll do
        is we'll have Frostie
        [flight engineer] oh,
        about a couple of
        minutes before touch-
        down, signal for a brace
        position.

Ftl Att.:   Okay, he'll come on
            the PA.

Capt.:   And then ah . . .

Flt Att.:   And if you don't
            want us to evacuate,
            what are you gonna
            say?

Capt.:   We'll either use the PA
         or we'll stand in the
         door and holler.

Flt Att.:   Okay, one or the
            other, we're reseating
            passengers right now
            and all the cabin lights
            are full up.

Capt.:   Okay.

Flt Att.:   Will go take it from
        there.
Capt.:   All right.
Flt Att.:   We're ready for your
        announcement any time.

### TIME 5:45:43 (seconds)

| INTRA-COCKPIT | AIR-GROUND COMMUNICATIONS |
| --- | --- |

Flt Eng.:   I can see the red
        indicators from here
        [trying to determine if
        the landing gear is
        down] but I can't tell if
        there's anything lined
        up.
1st Off.:   How much fuel we
        got, Frostie?
Flt Eng.: Five thousand
        [pounds].
1st Off.:   Okay.
Off duty Capt. sitting in cockpit
        hitching a ride . . . Less
        than three weeks, three
        weeks to retirement,
        you better get me outta
        here.
Capt.:   Thing to remember is,
        don't worry.
Off-duty Capt.:   Yeah, if I
        might make a sugges-
        tion, you should put
        your coats on . . . both
        for your protection and
        so you'll be noticed so
        they know who you are.

Capt.: Oh, that's okay.

Off-duty Capt.: But if it gets hot [a fire], it sure is nice to not have bare arms.

Capt.: But if anything goes wrong, you just charge back there and get your ass off, ok?

Off-duty Capt.: Yeah . . . I told the gal, put me where she wants me [to lend assistance] I think she wants me at a wing exit.

Capt.: Okay fine, thank you.

1st Off.: What's the fuel show now, Buddy?

Capt.: Five [thousand pounds] . . . figure about another fifteen minutes.

Flt Eng.: Fifteen minutes?

Capt.: Yeah.

Flt Eng.: Not enough—fifteen minutes is gonna— really run us low on fuel here.

Unknown voice: Right.

1st Off.: Think we ought to warn these people on the ground.

Capt.: Yeah, will do that right now.

| INTRA-COCKPIT | AIR-GROUND COMMUNICATIONS |
|---|---|
| **Capt.:** Call the ramp [at United terminal], give 'em our passenger count including laps [infants], tell 'em we'll land with about four thousand pounds of fuel and tell them to give that to the fire department. I want United mechanics to check the airplane after we stop, before we taxi. | |
| **Flt Eng.:** Yes sir. | |
| | **Flt Eng.:** Portland ramp, United one seventy-three. |
| | **United Ramp:** United one seventy-three, Portland, go [ahead]. |

Flt Eng.: United one seventy-three will be landing in a little bit and the information I'd like for you to pass on to the fire department for us. We have souls [people] on board, one seven two, one hundred and seventy-two plus five children . . . That would be five infants, that's one seventy-two plus five infants and pass it on to the fire department, we'll be landing with about four thousand pounds fuel and requesting as soon as we stop United mechanics meet the airplane for an inspection prior to taxiing further, go ahead.

United Ramp: One seventy three, copied it all and I'll relay that we're showing you at the field about zero five [landing about 6:05 PM], does that sound close?

Flt Eng.: He wants to know if we'll be landing above five after.

Capt.: Yeah.

Capt.:   You might—you might
         just take a walk back
         through the cabin and
         kinda see how things
         are going. Okay?

Flt Eng.:   Yeah, I'll see if
            its—get us ready.

1st Off.:   If we do indeed have
            to evacuate, assuming
            that none of us are
            incapacitated, you're
            going to take care of the
            shutdown, right? Park-
            ing brakes, spoilers and
            flaps, fuel shut off lev-
            els, fire handles, battery
            switch and all that.

Capt.:   You just haul ass back
         there and do whatever
         needs doing.

## TIME 6:01:34

| INTRA-COCKPIT | AIR-GROUND COMMUNICATIONS |
|---|---|
| Flt Eng.: [You've got] another two or three minutes. | |
| Capt.: Okay—how are the people? | |
| Flt Eng.: Well, they're pretty calm and cool, ah— some of 'em are obviously nervous—but for the most part they're taking it in stride. | |

Flt Eng.:   I, ah, stopped and reassured a couple of them, they seemed a little bit more—more anxious than some of the others.

Capt.:   Okay, well about two minutes before landing that will be about four miles out, just pick up the mike—the PA and say assume the brace position.

Flt Eng.:   Okay.

Flt Eng.:   We got about three [thousand pounds] on the fuel and that's it.

Port App.: United one seventy-three heavy, did you figure anything out yet about how much longer?

1st Off.:   Yeah, we, ah, have indication our gear is abnormal, it'll be our intention in about five minutes to land on [runway] two eight left, we would like the equipment standing by, our indication are the gear is down and locked, we've got our people prepared for an evacuation in the event that should become necessary.

Port App.:   Seventy-three heavy, okay advise when you'd like to begin your approach.

Capt.:   Very well, they've about finished in the cabin—I'd guess about another three, four, five minutes.

Port App.:   United one seven three heavy, if you could give me souls on board and amount of fuel.

Flt. Eng.:   One seventy two plus

Capt.:   One seven two and about four thousand, well, make it three thousand pounds of fuel.

Flt Eng.:   Plus six laps [infants].

Capt.:   Okay, and you can add to that one seventy-two, plus six laps, infants.

TIME 6:06:13

| INTRA-COCKPIT | AIR-GROUND COMMUNICATIONS |
|---|---|

(Sound of cabin door)
Capt.:   How you doing?
Flt Att.:   Well, I think we're ready.
Capt.:   Okay.

Flt Att.: We've reseated,
they've assigned helpers
and showed people how
to open exits.

Capt.: Okay.

Flt Att.: We have, they've
told me, they've got
able bodied men by the
windows.

Flt Att.: The [off-duty]
captain's in the very
first row of coach after
the galley.

Flt. Att.: He's going to take
that middle galley door,
it's not that far from the
window.

Capt.: Okay, we're going to
go in now, we should
be landing in about five
minutes.

1st Off.: I think you just lost
number four [engine]
buddy, you—

Flt Att.: Okay, I'll make the
five minute announce-
ment. I'll go, I'm sitting
down now.

1st Off.: Better get some
cross feeds [fuel lines
from other tanks] open
there or something.

Flt Eng.: Okay.

1st Off.: We're goin' to lose
an engine buddy.

Capt.: Why?

1st Off.:   We're losing an
               engine.
Capt.:   Why?
1st Off.:   Fuel.
1st Off.:   Open the crossfeeds,
               man.
Capt.:   Open the crossfeeds
               there or something.
Flt Eng.:   Showing fumes.
Capt.:   Showing a thousand or
               better.
1st Off.:   I don't think it's in
               there.
Flt Eng.:   Showing three
               thousand, isn't it?
1st Off.:   Its flamed out.

Capt.:   United one seven
               three, would like clear-
               ance for an approach
               into [runway] two eight
               left, now.
Port App.:   United one
               seventy three heavy, ok,
               roll out heading zero
               one zero—be a vector to
               the visual runway two
               eight left and ah, you
               can report when you
               have the airport in sight
               suitable for a visual
               approach.

TIME 6:07:27

INTRA-COCKPIT          AIR-GROUND
                        COMMUNICATIONS

Flt Eng.: We're going to lose
number three [engine]
in a minute, too.

Capt.: Well.

Flt Eng.: It's showing zero.

Capt.: You got a thousand
pounds, you got to.

Flt Eng.: Five thousand in
there, buddy, but we
lost it.

Capt.: All right.

Flt Eng.: Are you getting it
back.

1st Off.: No, number four,
you got that crossfeed
open?

Flt Eng.: No, I haven't got it
open, which one.

Capt.: Open em both, ****
[deleted] get some fuel
in there.

Capt.: Got some fuel pressure?

Flt Eng.: Yes, sir.

Capt.: Rotation now she's
coming.

Capt.: Okay, watch one and
two.

Capt.: We're showing down
to zero or a thousand.

Flt Eng.: Yeah.

Capt.: On number one.

Flt Eng.: Right.

1st Off.: Still not getting it.

Capt.: Well, open all four
crossfeeds.

Flt Eng.: All four?

Capt.:   Yeah.

1st Off.:   All right now, its coming.

1st Off.:   It's going to be ****
[deleted] on approach though.

Capt.:   You gotta keep em running, Frostie.

Flt Eng.:   Yes, sir.

1st Off.:   Get this ****
[deleted] on the ground.

Flt Eng.:   It's showing not very much more fuel.

Capt.:   United one seven three has got the field in sight now.

Port App.:   Okay, United one seventy-three heavy, maintain five thousand [feet].

Flt Eng.:   Number two is empty.

Capt.:   United, ah, one seven three is going to turn toward the airport and come on in.

Port App.:   Okay United one seventy-three heavy, turn left heading three six zero and verify you do have the airport in sight.

1st Off.:   We do have the airport in sight.

Capt.:   How far you show us from the field?

Port App.: I'd call it eighteen
flying miles.

Capt.: All right.

Flt Eng.: Boy, that fuel sure
went to hell all of a
sudden.

Capt.: There's ah, kind of an
interstate highway type
thing along that bank on
the river in case we're
short.

Unknown voice: Okay.

Capt.: That's Troutdale
[another airport] over
there, about six of one,
half a dozen of the
other.

1st Off.: Let's take the shor-
test route to the airport.

Capt.: What's our distance
now?

Port App.: Twelve flying
miles

Capt.: Okay.

Capt.: About three minutes—
four.

Flt Eng.: We've lost two
engines guys.

1st Off.: Sir?

Flt Eng.: We just lost two
engines, one and two.

1st Off.: You got all the
pumps on and every-
thing?

Flt Eng.: Yep.

Port App.:   United one
             seventy-three heavy,
             contact Portland tower
             [on frequency of] one
             one eight point seven,
             you're about eight or
             niner flying miles from
             the airport.

1st Off.:   Okay.

Capt.:   They're all going.

Capt.   We can't [even] make
        Troutdale [closer air-
        port].

1st Off.:   We can't make any-
            thing.

Capt.:   Okay, declare a May-
         day [emergency].

1st Off.:   Portland tower,
            United one seventy-
            three heavy, Mayday,
            the engines are flaming
            out, we're going down,
            we're not going to be
            able to make the air-
            port.

Tower:   United one. . . .

## IMPACT WITH TRANSMISSION LINES
(Heard on tower tape)

# Chapter 6

# Crew Distracted—
# Jumbo Jet Flies Into Ground

As you've read, the cause of the last two accidents was in the "unbelievable" category. One would think that it would be difficult to come up with another aircraft accident that was due to such improbable flight crew decisions. Unfortunately, that is not the case.

Let's look at another totally unexpected cause by moving our accident scene from the west coast of the United States over to the east. And, for this chapter, we'll show you how a full complement of flight crew members, consisting of a Captain, First Officer, Second Officer and a maintenance specialist, became so absorbed in repairing one tiny green light which failed to illuminate, that not one of them remembered to keep an eye on the instrument which indicated the altitude of the aircraft. Result? Their jumbo jet slowly, but inexorably, flew right into the ground.

A very small problem resulting in a very large disaster.

You are invited to come along as a silent observer in the cockpit so that you can see, step by step, how such a frightening event can actually take place.

This particular plane, a Lockheed 1011, was known as Eastern

Air Lines, Flight 401 (EAL 401). It had departed from New York's John F. Kennedy Airport at 9:20 PM on December 29, 1972, and was headed for Miami Airport (MIA) in Florida. There were a lot of people on board, 163 passengers and 13 crew members.

The flight was uneventful until the approach to Miami Airport. The landing gear handle was placed in the "down" position during the preparation for landing, and the green light, which would have indicated to the flight crew that the nose landing gear was fully extended and locked, failed to illuminate. The Captain recycled the landing gear, but the green light still failed to illuminate.

(For those of you not on board as observers, we will now describe the actions of the flight crew for the next crucial eight minutes. We will also report their conversations as recorded on the cockpit voice recorder. Explanations [in brackets] will be inserted for the purpose of clarification.)

At 11:34 PM, EAL 401 called the MIA tower and stated, "Ah, tower this is Eastern, four zero one, it looks like we're gonna have to circle, we don't have a light on our nose gear yet."

The tower advised, "Eastern four oh one heavy [jumbo jet], roger, pull up, climb straight ahead to two thousand [feet], go back to approach control [on frequency of] one twenty eight six."

The flight acknowledged, "Okay, going up to two thousand, [we will contact approach control on frequency of] one twenty eight six."

EAL 401 contacted MIA approach control and reported, "All right, approach control, Eastern four zero one, we're right over the airport here and climbing to two thousand feet, in fact, we've just reached two thousand feet and we've got to get a green light on our nose gear."

Approach control acknowledged the flight's transmission and instructed EAL 401 to maintain 2,000 feet and turn to a heading of 360 degrees. The new heading was acknowledged by EAL 401, which had no fuel problem and was only mildly concerned by the failure of the nose gear green light to illuminate.

The captain then instructed the First Officer, who was flying the aircraft, to engage the autopilot (which would automatically maintain the plane's altitude at 2000 feet). The First Officer acknowledged the instruction.

Since the aircraft was now being flown automatically by autopilot, the first officer was free to attempt to repair the nose gear light. He successfully removed the light lens assembly to examine, but it jammed when he attempted to replace it. When he continued to have difficulty with it, the Captain asked the second officer to descend into the electronics bay, located just below the cockpit, to visually check the landing gear linkage to determine if the nose gear, despite the absence of the green light, was actually extended.

Meanwhile, the controller, aware that the flight crew wanted to continue circling while they worked on the nose gear problem, radioed the flight to turn left to another heading. EAL acknowledge the request and turned to the new heading.

A few minutes later, the Second Officer climbed back into the cockpit and stated that it was too dark in the bay to make a determination of the nose gear. The crew then continued their attempts to free the nose gear light lens from its retainer, without success. Whereupon, the Captain again decided to send the Second Officer into the electronics bay to make another attempt to check the alignment of the nose gear. While this was going on, the autopilot was operating the controls and supposed to be maintaining the altitude of the aircraft at 2000 feet.

EAL 401 then called MIA approach control and said, "Eastern four oh one'll go out west just a little further if we can here and, ah, see if we can get this light to come on here." MIA approach control granted the request.

For the next two minutes, the Captain and the First Officer discussed the faulty nose gear position light assembly and how it might have been reinserted incorrectly. In the midst of that time period, a half-second musical C-chord, which indicated a deviation of 250 feet from the selected altitude, sounded in the cockpit. No crew member commented on the C-chord. No change in controls to correct for the loss of altitude was recorded. (The sounding of the C-chord must have been heard by the flight crew members since it was recorded on the Cockpit Voice Recorder.)

Although the autopilot was set to maintain the altitude of 2,000 feet, the aircraft had started to gradually descend and had now

gone below the 1,750 level. The audible one-half second musical
C-chord sounding a change in altitude of more than 250 feet from
the set gauges seems to have been ignored by the crew. Unfortunately, with so many warning systems installed in modern aircraft often sounding "alerts" in routine operations, there is a tendency to ignore or "not hear" another signal which may turn out
to be not routine, after all.

Shortly after, the Second Officer raised his head into the cockpit and stated, "I can't see it, it's pitch dark and I throw the little
light [in the electronics bay]—I get nothing."

The maintenance specialist, occupying the forward observer
seat, volunteered to descend into the electronics bay to assist the
Second Officer.

Meanwhile, Eastern 401, with the flight crew remaining
unaware, was continuing to lose altitude. Local weather, although
clear, with unrestricted visibility, was of no help as far as reference to the ground was concerned since it was close to midnight,
there was no moon, and the aircraft was flying over the desolate
Florida Everglades consisting of flat marshland. There were no
street lights, automobile headlights or lights from homes to offer a
clue to as the relative height above ground to any of the cockpit
crew members.

At this point, the Miami approach controller, looking at his
radar display, observed that the Eastern jumbo aircraft, instead of
indicating an altitude of 2000 feet, only showed 900 feet on his
screen. (He later testified that he had no doubt, at that moment,
about the safety of the airplane. Momentary deviations in altitude
information of the radar display, he said, are not uncommon, and
more than one scan would be required to verify a deviation
requiring controller action.) Noticing the discrepancy, the MIA
controller asked, "Eastern, ah, four oh one, how are things coming along out there?"

EAL 401 replied, "O.K., we'd like to turn around and come,
come back in [make its landing]." Approach control, without
mentioning the lower altitude reading on its radar, granted the
request with, "Eastern four oh one, turn left heading one eight
zero." EAL 401 acknowledged and started the turn (still losing
altitude) towards the airport.

At 11:42:05, while turning with the left wing lowered, the First Officer said, "We did something to the altitude." The captain's reply was, "What?"

The First Officer persisted, "We're still at two thousand, rights?" and the Captain immediately exclaimed, "Hey, what's happening here?" (At last, realization of the aircraft's shockingly low altitude.)

At 11:42:10, altimeter warning "beep" sounds began to ominously persist in the cockpit. The Captain and First Officer frantically tried to pour on power and pull back on the controls to arrest and reverse the descent of the aircraft. (Unfortunately, it takes four to six seconds for jet engines to "spoolup" and develop thrust.)

Two seconds later, with the aircraft in a left bank, Flight 401 ploughed into the Everglades swamp. The impact was recorded on the cockpit voice recorder at 11:42:12 PM.

(Both of the later recovered "Captain's and First Officer's altimeters indicated a reading of approximately 75 feet below sea level.)

The aircraft went down off the Tamiami Trail, a highway running through the Everglades from Miami to the west coast of Florida. The area, according to an Eastern Airlines spokesman was described as "inaccessible," consisting of flat marshland, covered with soft mud under six to 12 inches of water.

The left wing struck the ground first; the left engine, and then the left main landing gear, followed immediately. The aircraft disintegrated, scattering wreckage over an area approximately 1,600 feet long and 300 feet wide. The passenger compartment of the fuselage was broken into four main sections and numerous small pieces.

After impact, a flash fire developed from sprayed fuel. Some of the burning fuel penetrated the cabin area, causing many passengers to suffer burns on exposed body surfaces.

The search for the aircraft and the initial rescue efforts were coordinated by the United States Coast Guard, which was notified of the accident by Miami tower controllers. Helicopters were airborne almost immediately from the Coast Guard station at Opa Locka, Florida. The crash site was located about 15 to 20 minutes later.

The Coast Guard flew four doctors, three paramedics and nine corpsmen to the scene which could be reached only by helicopter and airboat. The highway patrol also dispatched a number of helicopters as well as a dozen ambulances; the latter standing by in the vicinity since they could not get to the crash site through the swampland. In addition, volunteer drivers of airboats participated in the rescue efforts.

A Coast Guard helicopter pilot observed from the air, "The plane is a mess. There are one or two or three large chunks. It is very dark. There are little pockets of people. There are bodies spread all around."

Two men on an airboat, hunting frogs in the Everglades, witnessed the accident. "We saw the plane go over real low," one said. "All of a sudden there was a bright flash that lasted 15 or 20 seconds." They reached the accident shortly and pulled victims from the shallow water. They later helped other rescuers give morphine to members of the flightcrew in the shattered cockpit which held them trapped in an upside-down position.

Since the crash occurred in the swamps, the absence of lighting of any kind at the scene seriously hampered survivors' ability to orient themselves and prevented them from searching for and assisting other injured survivors. Additionally, this lack of light prevented some of the surviving cabin attendants from taking effective charge among the passengers.

Rescuers who worked in the darkness were later aided by lights attached to helicopters and other aircraft circling overhead. In addition, generators were ferried in by helicopters and airboats to supply power to large lamps positioned on the ground in the crash area.

Some survivors had every particle of clothing stripped from their bodies but somehow had managed to escape serious injury. "The first person I saw was still sitting in his seat strapped in his safety belt with nothing around him but knee-deep water," said the pilot of a helicopter.

"He was talking to a girl sitting in the water next to his seat. They didn't pay any attention to us. It was like they were in such deep shock they were oblivious to us and the bodies lying in the mud around them."

An eerie sight was that of survivors who were able to walk, gathering in small clusters, some on dry ground, some in water up to their knees, singing Christmas carols in the darkness, while awaiting rescue. In the far distance, they could see the lights of the bustling Miami resorts.

None of the passengers had any indication that the airplane was about to crash. One of the survivors, a young man, stated that there was no word from the pilot, no warning, no explosion, nothing. "The plane was flying fine," he said. "We just went down slightly . . . I thought nothing of it. The next thing I knew we hit."

Among the rescuers were Miccosukee Indians who lived in the swamp area. They helped with the rescue work but refused to allow the dead to be laid out in their schoolhouse. This was because their customs would require them to tear the schoolhouse building down afterward.

Along with the heartwarming and unselfish efforts of the varied rescuers, there was one note of human misbehavior to mar the scene. Looters were seen taking jewelry and wallets from bodies around the disaster area. "While we were working out there, a lot of us noticed the looters," said a Coast Guardsman. "I saw them taking watches and things from dead people and so did some of the game wardens and others. But what can you do? We were trying to help the survivors and get them out of there," he said.

The rescue operation lasted until the dawn hours. By that time, all the injured had been transported to Miami hospitals by the Coast Guard, Army and Air Force helicopters.

About 100 friends and relatives who were awaiting the arrival of the plane at Miami Airport, were directed to an airport lounge by Eastern officials. They were told at 12:30 AM that the aircraft had crashed. In the hours afterwards, as word of survivors kept coming in, these people dashed from hospital to hospital in hopes of finding their loves ones.

"It's a wonder anyone got out alive," said the director of the Miami Port Authority. He had spent the night, along with other rescuers, searching for victims in the waist-deep alligator holes and snake-infested swamp around the perimeter of the wreckage.

An equal number of relatives of passengers aboard the jetliner were flown in by Eastern from New York. They were met by airline officials who took them to the hospitals where survivors had been taken. The airline also provided hotel accommodations for any of the relatives who needed them.

Relatives of those known to have died in the crash were notified by Eastern that they would make arrangements to take the bodies to places designated by the next of kin.

It took several days before a final tally could be made of the number of survivors and those fatally injured.

### Official Count of Injuries to Crew and Passengers

| INJURIES | CREW | PASSENGERS | TOTAL |
|----------|------|------------|-------|
| FATAL | 5 | 96 | 101 |
| NONFATAL | 8 | 67 | 75 |

Three members of the flight crew, as well as two of the flight attendants, died in this accident. Eight of the 10 stewardesses survived.

Before an official investigation got under way, an Eastern pilot and spokesman stated that under normal circumstances, Flight 401 should have been maintaining an altitude of 2,000 feet and cruising at a speed of 225 miles an hour at the time it struck the ground and pancaked across the flat swampland of the Everglades. Although the fuselage was shredded away as it went through the mud and sowgrass of the swamp, the angle of descent spared a more traumatic impact. In addition, the large cargo compartments beneath the passenger deck also provided a crumpled buffer between the passengers and the ground.

Based on the wreckage path and furrows in the mud, some air experts believed that the plane landed in a relatively straight level, the left wing touching first and digging a long trough through the mud. Their unofficial hypothesis was that the flight crew, absorbed in trying to fix a nose-gear problem, failed to differentiate the swampland from the nighttime horizon. With the airport lights over his left shoulder, the pilot would have seen only the black midnight sky and below it, with no hint of separation, the dark swamp. (Two commercial pilots who later flew newsmen to the swamp crash site before dawn wanted to turn back because they had no visual horizon to orient themselves with the ground.)

However, the 1011 aircraft actually has two altimeters, which tell the flight crew exactly how far above the ground they are at all times. The problem was, no one was paying attention to these crucial instruments—until just before impact, which, of course, was too late.

The National Transportation Safety Board received notification of the accident shortly after midnight and immediately dispatched a team to the scene.

The thrust of the investigation was focused on ascertaining the reasons for the unexpected descent. One theory considered was the possible incapacitation of the pilot.

A post-mortem examination of the Captain revealed that he had a moderate-sized tumor on the right side of the brain. This discovery led to a number of possibilities that the investigating board pursued.

The medical examiner suggested that this tumor could have affected the Captain's vision particularly where peripheral vision was concerned. Additionally, expert testimony revealed that the onset of this type of tumor is slow enough to allow an individual to adapt, by compensation, to the lack of peripheral vision so that neither he nor other close associates would be aware of any changed behavior.

It was hypothesized that if the Captain's peripheral vision was severely impaired, he might not have detected movements in the altimeter while he watched the First Officer remove and replace the nose gear light lens. However, the Captain's family, close friends and fellow pilots advised that he showed no signs of visual difficulties in the performance of his duties and in other activities requiring peripheral vision.

Based on the testimony of these people who would have been in a position to detect any aberration in the Captain's normal behavior, and in the absence of any indications to the contrary, the Safety Board ultimately came to the conclusion that this tumor was not a factor in this accident.

Another area considered by the investigating committee as a cause of the aircraft flying into the ground was the autoflight system.

It was noted that the First Officer flew the airplane manually away from the airport after the discovery of the malfunctioning nose gear light. The Captain then ordered the engagement of the autopilot to maintain 2,000 feet of altitude so that the flight crew could concentrate on the light.

The recovered Flight Data Recorder (sometimes referred to as the "black box" even though it is painted orange-red for easier identification) indicated that five minutes before impact there was a vertical movement causing a 200 feet-per-minute rate of descent. For a pilot to induce this movement, he would have to disengage the altitude hold function. It is conceivable that this could have been produced by an inadvertent action on the part of the pilot which caused a force to be applied to the control wheel sufficient to disengage the altitude hold position. It was noted that this occurred at the same time the Captain commented to the Second Officer to "Get down there and see if the . . . nose wheel's down." If the Captain had applied a force to the control wheel while turning to talk to the Second Officer, the altitude hold function might have been accidentally disengaged.

To make sure there was no mechanical malfunction of the autopilot system, it was necessary to examine the wreckage, which included the flight instruments. At the conclusion of the examination, the Safety Board ruled out any deficiencies in the autoflight system as a critical factor in the accident.

One other possible cause of the unintentional descent to be investigated, was the performance of the flight crew.

In addition to the disengagement of the altitude hold function set at 2,000 feet, a series of reductions in power began less than three minutes before impact, which could also cause the airplane to descend. The Captain might have inadvertently bumped the throttles with his right arm when he leaned over the control pedestal to assist the First Officer, or the First Officer's left arm might have accidentally bumped the throttles while he was occupied with the nose gear indicating system. The other alternative is that one of the pilots intentionally reduced thrust power when he noted that the speed of the aircraft was exceeding the desired speed for circling.

Regardless of the manner in which the thrust reduction occurred, the flight instruments would have indicated abnormally for a level-flight condition. Together with the altitude-alerting C-chord signal, the flight instrument indications should have alerted the crew to the undesired descent.

The Safety Board stated that it is aware of the distractions that can interrupt the routine of flight. However, they felt that once the autopilot was engaged to fly the aircraft and reduce the workload in the cockpit, one of the flight crew should have been delegated to keeping an eye on the flight instruments.

After a six-month investigation, the Safety Board came to the conclusion that, "The probable cause of this accident was the failure of the flight crew to monitor the flight instruments during the final four minutes of flight, and to detect an unexpected descent soon enough to prevent impact with the ground. Preoccupation with a malfunction of the nose landing gear position indicating system distracted the crew's attention from the instruments and allowed the descent to go unnoticed."

An interesting postscript of the accident of this Lockheed 1011 aircraft focused on the relatively high survival rate of the cabin occupants. There seemed to be something new to be learned about safety involving passengers unlucky enough to experience an airplane crash.

The Lockheed 1011, known as the Tristar, is a three-engine jetliner. It was a relatively new commercial plane having made its maiden flight approximately two years earlier (in 1970). This was the first crash of this aircraft. As a matter of fact, this was the first major accident (at the time) involving any of the new generation of wide-bodied aircraft or jumbo jets, which included the Boeing 747 and the Douglas DC-10.

It is remarkable that, despite the fact the airplane flew into the ground, a number of passengers walked away from the crash with only minor injuries. Since this was the first fatal accident of any jumbo jet, investigators initially expressed the hope that the high survival rate was due to an extra measure of protection against impact forces built into the large frame of this type of aircraft. (After investigating a number of wide-bodied accidents that took place in later years, this theory was discarded.)

What did develop from the investigation was that although the fuselage shell was torn away exposing the occupants to external hazards, the fuselage structure itself did not impinge on the survivors. Seventy-five people survived because either their seats remained attached to large floor sections or the occupants were thrown clear of the wreckage at considerably reduced velocities.

The major survival factor was the design of the passenger seats in this aircraft. These seats incorporated energy absorbers in the support structure. In addition, in contrast with the conventional floor tiedown arrangement of aircraft seats, each of these seats was bolted to a platform, which in turn, was attached to the basic aircraft structure. It was noted with great interest that many of the seat units remained attached to these platforms. There was no question that the design of the seats materially contributed to the survival of many occupants.

As we all know, new commercial jet aircraft take years to design and test before being accepted by airline companies to fly their precious passengers. We can only hope that the high survival rate of the people in this accident due to the unusual seat design, will be remembered and taken into consideration when plans are finalized for any new generation of passenger aircraft.

(Note: A five year research and development program on the design of seats was conducted by the Federal Aviation Administration in collaboration with the National Aeronautics and Space Administration. Based on the results, the FAA adopted a rule requiring airliners coming off the assembly line to be installed with stronger seats that would be less apt to tear loose from the floor. In May, 1988, the FAA went one step further. In addition to the adopted rule affecting new passenger aircraft, the agency proposed a rule requiring all existing airplanes to be re-fitted with the new stronger seats within seven years of the adoption of this new rule.)

Tires blow out on takeoff. A part of the wing of a Continental DC-10 is shown in front of the charred hulk of the aircraft at Los Angeles International Airport, March 1, 1978. (AP/Wide World Photos)

In a hurry. Wreckage of KLM Boeing 747 jetliner litters field at Tenerife, Canary Islands, March 27, 1977. (AP/Wide World Photos)

Collision in the air. A Pacific Southwest Airlines jetliner plummets to earth over San Diego, September 25, 1978. (AP/Wide World Photos)

San Diego neighborhood where jetliner crashed, leaving a trail of destruction. In foreground is where the plane finally stopped. (AP/Wide World Photos)

Landed on the wrong runway. A part of the cabin, the largest piece of wreckage from Western Airlines DC-10 that crashed in Mexico City, October 31, 1979. (AP/Wide world Photos)

Ran out of fuel. The twisted wreckage of a United Airlines DC-8 lies among the trees in a residential Portland, Oregon, neighborhood, December 28, 1978. (AP/Wide World Photos)

Crew distracted--jumbo jet flies into ground. The remains of the rear jet engine and part of tail assembly of Eastern Air Lines Lockheed 1011 that crashed in the Everglades, December 29, 1972. (AP/Wide World Photos)

Cargo door opens in flight. Firemen and rescuers search the smoking debris of Turkish Airlines DC-10 that crashed into forest north of Paris, March 3, 1974. (AP/Wide World Photos)

Wind shear on landing. Wreckage of Eastern Air Lines jet is in foreground as a DC-8 jetliner prepares to land at Kennedy Airport, New York, June, 24, 1975. (AP/Wide World Photos)

Wind shear on take-off. A fireman stands on wing of Pan Am Boeing 727 that crashed into a Kenner, Louisiana, neighborhood, July 9, 1982. (AP/Wide World Photos)

Both engines flame out in thunderstorm. A Southern Airways DC-9 crashed while attempting to land on a roadway after its engines failed., April 4, 1977. (AP/Wide World Photos)

On a collision course with the tallest skyscraper. Wreckage of a B-25 bomber protrudes from the side of the Empire State Building, July 28, 1945. (AP/Wide World Photos)

Engine falls off on take-off. Aftermath of the crash of an American Airlines DC-10 after taking off from Chicago-O'Hare International Airport, May 25, 1979. (AP/Wide World Photos)

Strayed off course--attacked by fighter aircraft. Soviet naval officers spread out pieces of Korean airliner downed by Russian fighters on September 1, 1983. (AP/Wide World Photos)

Snow and ice on the wings. A U.S. Park Police helicopter pulls two people from the wreckage of an Air Florida jetliner that fell into the Potomac River, January 13, 1982. (AP/Wide World Photos)

Gunfire in the cockpit. Investigators search the impact area where a Pacific Southwest jetliner crashed on December 7, 1987. (AP/Wide World Photos)

# Chapter 7

# Cargo Door Opens in Flight

As we pointed out before, the causes of accidents are different and unique. In the last accident that we discussed, a gear light that failed to illuminate initiated a series of decisions that ultimately led to the destruction of the Lockheed jumbo jet. As the Safety Board indicated, the crew in the cockpit became distracted.

We will now focus on the performance, not of the flight crew, but of the ground crew working on a different jumbo airplane, our familiar DC-10. In this situation, the action takes place in Europe, at a French airport. And the culprit, this time, is the latch on a small cargo door on this multi-million dollar aircraft. Again, we don't want to get ahead of the story so we'll start with the giant airliner leaving Turkey on the first leg of its flight to France.

On Sunday, March 3, 1974, a Turkish Airlines DC-10 departed Istanbul and landed at Orly Airport in Paris at 11:00 AM local time, on schedule. There were 167 passengers on board Flight TK 981, of whom 50 disembarked. One hundred seventeen passengers remained on the jumbo jet for the next and final destination, London.

The normal stop is for one hour but was increased an extra half-hour because of the last minute embarkation of numerous passengers from British Airways and Air France, a large number of them British and Japanese nationals. A sudden strike at London's Heathrow Airport had forced the cancellation of many flights whose passengers switched to the Turkish airliner. These fresh passengers numbered 217 and boarded after passing through the routine police checks. The wide-bodied jet was now filled to capacity.

The usual maintenance procedures took place during the one-and-a-half-hour stop at Orly. The aircraft was refueled, passengers disembarked, freight and baggage removed, new passengers guided onto the aircraft, new cargo and baggage loaded into the cargo compartments, etc., etc. There seemed to be only 1 minor hitch in the routine—that of the closing and latching of the aft cargo door located at the left-hand side of the airplane.

The aft cargo compartment was closed by a cargo-handling operator who later stated that the door was closed without any particular difficulty and that he did not notice any abnormality. However, although the handle had been pulled down and the door closed, the lock pins on the door were not engaged and no visual inspection had been made through the small viewing window provided for the sole purpose of verifying that the lock pins were in place.

In an investigation afterwards, the cargo operator testified that he did not look through the view port, a procedure which he had seen but which he never carried out himself and the purpose of which he did not know. Nor did anyone see the ground engineer (a major part of his duties consisting of supervising loading and unloading) or any other crew member inspect the lock pins of the cargo door to verify engagement by looking through the view port.

Finally, the DC-10 jumbo jet was ready to depart. It was cleared to taxi to the runway in preparation for take-off. Meteorological conditions, including visibility, were good.

Aboard the Turkish airliner with its London destination, were 3 flight crew members, 1 ground engineer, 8 flight attendants and 334 passengers, including 6 children and 1 infant. That came to a total of 346 people on this flight. The takeoff weight of the heavily-loaded aircraft was approximately 360,000 pounds.

Flight 981 started its takeoff at 12:30 PM. Its ground roll time was 40 seconds and it smoothly lifted off the runway at 164 miles per hour. The climb-out progressed normally. The aircraft reached 6,000 feet, remained at that level for two minutes and then resumed its climb at a speed of 345 miles per hour.

Ten minutes after takeoff, just before reaching 12,000 feet, with the altitude in the cabin of the aircraft maintained at ground level for the comfort of the passengers, the differential pressure between the outside air and that in the fuselage reached the force of approximately five pounds per square inch. This was normal for that altitude. However, because of this pressure, a startling chain of events was set in motion.

All of a sudden, the two bolts on the aft cargo door snapped off. This was followed instantaneously by the opening of the latches, the lock pins of which had never properly been engaged.

Up to this point, as long as the two bolts held, the door structure withstood the mounting strain of the increasing pressurization force from the climb. When the two bolts gave way, the latches opened and the door itself sprung open, broke into several pieces, and fell away from the aircraft. Now, instead of a closed cargo door, there was a big hole in the fuselage.

This opening caused an immediate drop in the pressurization in the cargo compartment beneath the passenger cabin floor. The pressure relief vents located between the cargo compartment and the passenger cabin (purpose—to reduce the effects of rapid depressurization) were not large enough to accommodate the enormous discharge of air which rushed through the opened cargo door. As a result, this caused an instantaneous tremendous excess pressure above the floor.

This pressure created havoc with the aircraft and its passengers. As the noise in the cabin became deafening, the extreme pressure differential had a devastating effect upon the passengers seated in the rear of the airplane. The pressure was so incredibly powerful, that six people, still strapped into two triple-seat units located above the cargo door, were literally sucked out of the opening—together with their seats—and ejected into the atmosphere more than two miles above the ground. As they tumbled uncontrollably towards the earth at a descent of 12,000 feet per minute, it isn't difficult to sense the agonizing terror experienced by the six helpless passengers contemplating their impending death during that 60-second eternal, free-fall.

The remaining passengers were stunned and in shock. Many of them started to scream as pieces of seat wreckage, cushions, pillows, magazines, and anything in the rear of the aircraft that was not tied down, were pulled out of the opening and propelled into space.

The noise of decompression was heard on the cockpit voice recording. Also heard was the co-pilot saying, "The fuselage has burst," followed by the sounding of the pressurization aural warning.

The DC-10 was seriously wounded. All the horizontal stabilizer and elevator control cables run beneath the floor of the aircraft from the cockpit to the tail surfaces. The drop in pressure in the cargo compartment was severe enough to buckle and shatter the floor structure, and impair the flight control cables located under the floor. Now, controlling the jumbo jet became impossible.

Immediately after decompression, the nose of the DC-10 started to point down and its descent became rapidly steeper. Within seconds, the speed of the aircraft increased from 345 to 416 miles per hour even though two of its three engines had been throttled back by the pilot and the third one lost power on its own (that engine was located in the rear and its controls were also affected by the jammed or ruptured cables under the floor). By the time the airplane descended to 7,200 feet, the speed was up to 460 miles per hour.

The controller, who was following the progress of Flight TK 981 and had heard the pressurization warning on his radio, now heard the overspeed warning. (The DC-10 was still descending nose-down and the airspeed had now reached almost 500 miles per hour.)

At the same time as the overspeed warning signal was heard, the label with the flight number "981" disappeared from the secondary radar scope of the controller. On the primary radar, the aircraft echo (image) split in two. One part (displaying objects ejected from the aircraft) remained stationary for several minutes before fading out, while the second part, the echo of the DC-10 itself, continued on a path which curved away to the left.

The jumbo jet, descending at an extremely high speed, initially skimmed the tree tops of a forest and then, continuing its downward flight, smashed and destroyed hundreds of trees, carving out a path over 1,300 feet long and 300 feet wide, before hitting the ground with a tremendous explosion. The impact was so violent that the aircraft literally disintegrated into hundreds of thousands of pieces of fragmented wreckage. There was no fire because of this disintegration at very high speed.

The crash took place in the forest of Ermenonville, located at a place called Le Bosquet de Dammartin. The time was 12:42 PM; just about 70 seconds after the cargo door was blown off.

There were no survivors. All of the 346 persons on board Flight TK 981, which included 23 Americans, perished. (Ironically, this was the first fatal crash of a DC-10 in its two-and-a-half years of commercial service.)

The Air Traffic Control, immediately aware of the loss of radio and radar contact, was able to locate the area of the accident, thereby simplifying the task of the search services. Within a very short time, rescue teams arrived at the site.

It was a strange and desolate scene that greeted the would-be rescuers. One newspaper reported that the ground "was smothered with a chaos of debris and shards of metal." Another stated that the area "was transformed into a forest of ghastly Maypoles, the tops of trees sheared off and the spars fluttering with streams of clothing, paper, wire, shredded seats, dangling shoes and human limbs."

Since there was no one left alive to rescue, the searchers, in blue and white uniforms, poked and stirred the ruins with splintered branches looking for bodies, passports, jewelry, clothing and anything that could be used for identification. Bits of remains were collected and put into blankets and bags and placed on stretchers.

The dismembered bodies of the victims were brought to an old church, Saint-Pierre at Senlis, where they were laid out in the gothic, vaulted hall which served as a temporary morgue.

Seventeen emergency organizations, both civil and military, with 56 vehicles of various kinds (ambulances, fire engines, police cars, military trucks, etc.), were employed in the operations.

The disaster attracted thousands of people who clogged the roads with their automobiles or arrived on foot, many carrying children in their arms or tugging at dogs at the end of a leash. To maintain order, it was necessary to call out, in addition to policemen, an army of troops to open the roads for the siren-braying vehicles and keep the curious away from the wreckage site.

The following morning, French experts, accompanied by police officers from the area of Saint Pathus, searched the terrain beneath the flight path of the DC-10, looking for clues as to the reason for the crash of the airliner. Approximately nine miles from the main wreckage site, they found the cargo door. Also, much to their amazement, they discovered the bodies of the six passengers, clustered together and relatively intact, about two miles from the cargo door, with parts of the seats found near the bodies. The six newly-discovered victims were taken to Meaux Hospital.

On March 4, 1974, a Commission of Inquiry was appointed by the Minister of Transport of France to investigate the accident. The members of the Commission were assisted by French, Turkish, and Swiss experts as well as representatives from the U.S. National Transportation Safety Board, Federal Aviation Administration and McDonnell Douglas, the manufacturer of the DC-10. (A spokesman for the Safety Board in Washington said, "We want to be on hand to know as rapidly as possible what happened. I would say more than 100 DC-10's are used by the U.S. fleet. If there is a malfunction or a fault in the design, we want to be able to take corrective measures as fast as possible.") In addition, British and Japanese accredited observers were authorized to follow the course of the investigations, since the passengers included British and Japanese nationals.

As part of the Inquiry, it was decided to take the remains of the passengers and crew members to the Institut Médico-Légal de Paris to obtain medical and pathological information. In view of the exceptionally large number of victims, the medical team encountered difficulties, as the Institut did not have all the facilities to contend with such as extraordinary disaster. Nevertheless, they performed admirably.

The results of the examination of the victims were as follows:

a)  In the case of the bodies recovered at the main accident site in the forest, there was a high degree of fragmentation associated with the violence of the impact. This was so extensive that nearly 20,000 fragments of human remains were recovered and listed—an average of more than 59 fragments per victim. Completely solving this gigantic human jigsaw puzzle would obviously turn out to be impossible.

b)  The 6 bodies found near Saint-Pathus revealed major fractures and serious internal injuries. They, at least, individually, were all in one piece.

Careful examination of these six passengers showed that there were no external burns and no external wounds which could be associated with the projection of metal or other fragments as the result of an explosion. In addition, there was no evidence of deep penetration by metal fragments. Thus, the Commission could rule out the possibility of sabotage or an explosion caused by a criminal act. (A wire service had earlier reported Turkish airline sources speculating that five of the passengers might have been guerrillas carrying bombs that exploded in flight.)

Identifying the six ejected passengers was not too difficult. However, attempting to identify the remaining 340 victims proved to be a formidable task.

Due to the severe disintegration of the aircraft, the identity of almost half of the occupants of the ill-fated jumbo airliner could not be established. On the other hand, it was a remarkable achievement that as many as 188 bodies (or parts of bodies) were positively identified. This was accomplished by the use of a number of techniques which included fingerprinting (in particular in the case of Turkish and Japanese nationals because of the existence of national fingerprint records), examination of teeth, measuring of bones and the sorting out of clothing and personal effects.

Fingerprinting turned out to be of greatest assistance. What also turned out to be helpful was the use of a computer for processing the enormous quantity of data required for identification purposes.

While these human clues were being sorted out, another major problem facing the authorities was discovering the cause of this unusual air catastrophe.

The Commission of Inquiry arranged for the collection and transfer of all of the pieces of the aircraft wreckage to a hangar where they could be assembled and examined. This detailed operation started five days after the crash and took another 12 days to complete.

The aircraft's mechanisms and structure were examined carefully. It did not take too long for the experts to zero in on the culprit, the sudden opening of the aft left-hand side cargo door which caused the depressurization with its subsequent jamming of the controls of the jet. Especially since there was a history of similar circumstances in a previous accident to a DC-10 less than two years earlier.

An American Airlines DC-10 had taken off from Detroit, Michigan, on June 12, 1972, and was climbing over Windsor, Ontario, at just under 12,000 feet (same altitude as the Turkish Airlines DC-10) when the aft left-hand side cargo compartment door (also the same) blew out causing the cabin floor to collapse, severely damaging the control cables from the cockpit to the tail. (A coffin was sucked out of the open cargo door and was found in a field the next day with the body inside.) In this case, the pilot managed to land safety back in Detroit with the rear engine dead and without brakes or rudder.

The National Transportation Safety Board, investigating this Windsor incident, came up with two recommendations more than 20 months before the Turkish Airline crash. It said that the FAA (Federal Aviation Administration) should require a foolproof cargo door-locking mechanism and additional venting in the passenger floor to prevent collapse from decompression.

Both the FAA and McDonnell Douglas, the DC-10 manufacturer, were concerned about what had happened to the American Airlines DC-10 over Windsor, Ontario, and determined to prevent other such incidents. However, there was a difference of opinion between McDonnell Douglas and the FAA engineers as to the extent of the safety modifications that should be made.

The discussions between the FAA and McDonnell Douglas involved three possible steps that could be taken:

1. The installation of a small viewing window on the cargo door so that ground personnel could actually see whether the lock pins were engaged.

The FAA and the manufacturer were in agreement on this safety feature. Within a month after the close call of the jumbo airliner over Windsor, McDonnell Douglas notified all airlines employing DC-10's of the view-window change in an "alert" Service Bulletin, a notice on blue paper reserved for important safety-related items.

2. The installation of a modification to the locking mechanism on the cargo door which would make the door latching mechanism safe regardless of human error or abuse of the mechanism.

This change also received FAA approval. However, McDonnell Douglas did not believe that the urgency of this notice was as great as the window installation. Therefore, this Service Bulletin was issued on white paper saying that compliance was "recommended," which placed it in the same category of importance that a change in lighting fixtures or a rearrangement of seats would get. Some planes weren't modified under this notice for nearly a year and one wasn't modified until after the Paris crash, almost two years later. And, as a matter of fact, even a spokesman for the manufacturer, referring to the Turkish Airline DC-10 crash, stated that "preliminary evidence indicates the aft bulk cargo door on that particular aircraft didn't incorporate all the approved changes."

3. The installation of additional venting to prevent collapse of the floor due to sudden compression.

Although the other two modifications proposed by McDonnell Douglas were approved by the FAA and Service Bulletins issued almost immediately, this third safety measure proved to be considerably more difficult to reach agreement that it was absolutely necessary.

The manufacturer felt strongly that the new type of latching mechanism, plus the ability of now being able to visually observe (through the new port window) that the lock pins were properly engaged, was more than adequate to assure the fact that the cargo door was properly closed and would remain closed. Therefore, the danger of decompression had been eliminated and it was unnecessary to incur the expense of $250,000 to $500,000 per DC-10 for the refitting and installation of additional venting.

Nevertheless, the FAA was not convinced. There ensued an exchange of letters over a two-year period, starting with the time of the Windsor incident, in which the FAA kept raising questions about the safety problem involving decompression in all wide-body jet aircraft, including the Boeing 747 and Lockheed L-1011 as well as the DC-10. The answers they received from McDonnel Douglas still did not satisfy the FAA.

In one letter, an FAA engineer wrote the manufacturer stating that the loss of any airplane because the floor collapsed after the loss of a cargo door and resulting decompression would be "unacceptable." (This letter was written one year before the Turkish Airline crash.) In a later letter he asked the company for "your views on explosive decompression effects" and sought an analysis proving that the cargo door blowout was extremely improbable.

The manufacturer replied that questions raised in the FAA letter needed "much clarification" and pleaded that the company didn't have the available manpower to turn out such a report. McDonnell Douglas stated that it wasn't "in a position to accept this burden alone." The Company suggested that the FAA itself fund the study because, it contended, the results also would apply to the jumbo jets manufactured by other companies.

That last letter from the manufacturer was dated six days before the Turkish Airlines crash in Paris.

What was tragically prophetic was another document, a memorandum written by an engineer for the company who built the doors for McDonnell Douglas (a spokesman for the latter denied the memo was ever sent to them). This engineer strongly believed that the safety modifications made after the Windsor incident amounted to "a Band-Aid fix." He stated that what should have been done instead was to strengthen the cabin floor and to install blowout panels in it to guard against collapse.

His memo went on to say, "It seems to me inevitable that, in the 20 years ahead of us, DC-10 cargo doors will come open and cargo compartments will experience decompression . . . and I expect this to usually result in the loss of the airplane." The floor changes would be costly, he noted, but "may well be less expensive than the cost of damages resulting from the loss of one planeload of people."

His prophecy came true. Within a month after the crash of the Turkish Airlines DC-10, a class action suit asking more than $125 million in damages was brought against McDonnell Douglas and filed in federal court in the United States on behalf of the widow of one of the victims of the crash.

At a press conference later, the president of McDonnell Douglas insisted that the company was always conscientious concerning the safety of its aircraft and pointed out the fact that, after the Windsor incident, modifications in all DC-10s in service were made or recommended almost immediately. "No one can tell me that was slow action. That was dynamic action," he stressed.

He went on to say, "If the door [on the Turkish Airlines plane] had been properly shut, latched and locked, and the visual check had been made that it was locked, the tragic accident wouldn't have occurred." He termed it "inexcusable" that the baggage handler reportedly in charge of closing the cargo door was "untrained" and unable to read the directions in English. He stated that placards using international sign language describing the closing of the door are being passed out to airlines.

At the same press conference, in answer to a question, the chairman of the company admitted that "no new bookings" have been received for the DC-10 since the accident. He added that they had "one or two sales we had been working on and had hoped to get" but those airlines chose airplanes of competitors instead.

The Turkish Airlines disaster managed to bring the manufacturer and the FAA into complete agreement that the current safeguards against depressurization were inadequate. As a precautionary measure, McDonnell Douglas immediately advised all operators of DC-10 aircraft to review procedures and insure that previously issued service instructions relating to cargo door latching were being strictly observed. The FAA, on its part, sent telegrams to all airlines using the airplane, ordering flight crews to check each cargo door before takeoff.

Finally, the Federal Aviation Administration issued an order that all wide-bodied jets be modified so that a major puncturing of the fuselage would not be likely to lead to a crash. The requirement would be that these airplanes would be fitted with a venting system enabling them to withstand decompression from a hole as large as 20 square feet in the fuselage. The system would serve to bleed off the pressure in the airplane when one location, such as the cargo hold, experiences sudden decompression and threatens explosive depressurization in the rest of the plane.

This directive would apply to both new and existing aircraft, including the DC-10, Boeing 747 and Lockheed L-1011. It was expected that this venting system would be adopted by foreign airlines as well as the 275 wide-bodied jets being flown by U.S. Airlines at the time.

The Windsor incident took place on June 12, 1972; the Turkish Airlines crash on March 3, 1974. The new venting system was required to be installed on all jumbo jets no later than July 1, 1977. At last, the danger of explosive depressurization on all jumbo aircraft, was eliminated.

## Chapter 8

# Help Me—I Can't Fly This Airplane

The Turkish Airlines incident set another regrettable record—346 deaths from the crash of one giant passenger jet aircraft.

We'll now turn to a frightening episode at the other end of the scale—one involving a tiny propeller airplane carrying only one pilot and one passenger. Obviously, not to be compared with the Turkish Air disaster, but frightening, nevertheless.

"Help me. Help me. I can't fly this airplane. Please—anyone—help me."

This was the terrified cry of a 40-year-old woman coming over the airwaves on an aircraft radio frequency. She was seated in a small single-engine Piper Warrior airplane, 3,500 feet up in the air and, not being a pilot, had no idea how to control the airplane to bring it down to a safe landing.

Her husband, age 41, who had been handling the controls only moments ago, had suddenly slumped over with a heart attack leaving her in a perilous position. Having just experienced a personal tragedy in the unexpected death of her husband of 23 years, she was now faced with the likely prospect of personally becoming the victim of a second tragedy.

This drama was taking place on February 22, 1981. It had been

a beautiful Sunday afternoon and her husband had suggested they
rent an airplane that day (he had received his flying license 13
years ago) for a few hours and fly to a small city 80 miles away,
have a snack, and return. They would enjoy the trip and the scen-
ery enroute, something they had done with reasonable frequency
both with and without their two children.

To rent an airplane, they followed their usual procedure. They
called ahead to have a Warrior single-engine airplane serviced
and waiting for them, took off from Mocksville Airport in North
Carolina and flew into Morgantown, a small airport less than an
hour away. After strolling around for a short while and stretching
their legs, they climbed into the airplane for the return trip.

On the way back, cruising along at about 140 miles an hour,
her husband remarked that he felt somewhat faint and, in an
instant, collapsed in his seat. Alarmed, she turned to offer him
assistance and saw his face turn completely white and his eyes
roll back in his head. At that moment, panic set in—she knew she
had lost the man who had meant so much to her and, at the same
time, she was aware that she was now the only living person at
the controls of the aircraft. (Small airplanes, like the Piper
Warrior, generally have dual controls in the front, one for the
pilot, who can sometimes act as an instructor, and the other for
the front seat passenger, who can be, at times, a student or a co-
pilot.)

Despite the fact that she had flown with her husband a number
of times, she had never learned to fly although her husband had
tried to interest her in that skill. It was one of those situations
where she felt it might be too difficult to master the controls and
instruments. Besides, she had the utmost faith in the ability of her
husband to take her safely up and bring her safely down. Her
enjoyment was being together with him, looking out the front and
side windows and occasionally taking pictures. Now, she was
faced with a situation that, in her wildest dreams, she could never
imagine herself being in.

"Help me. Help me," was the cry that came over the airwaves.
A cry for her husband and a cry for herself.

Although she did not know how to fly the airplane, she had
seen her husband turn the control wheel, like the wheel of an

automobile, in order to turn the airplane to the left or to the right. But which way was the airport? To an inexperienced person looking down from 3,500 feet, it is extremely difficult to distinguish one landmark from another, with the exception, perhaps, of a body of water or a major highway. Even if she were fortunate enough to know where the airport was located and maneuver the airplane in that direction, what levers, knobs, controls and pedals would she use to safely place the airplane on the runway?

However, at least the radio seemed somewhat familiar. It resembled the CB radio she had in her car. She pressed the transmitter button on the aircraft's microphone and begged somebody, anybody, for help.

Fortunately, the frequency that the radio was tuned to was one used by many pilots for air-to-ground communications. Her cry of distress was heard by the Atlanta Center controller, and picked up in cockpits of other aircraft as well as a number of different airport communications centers, all tuned to that frequency.

When they heard the frantic voice explain that her husband had been stricken and that she did not know how to fly the airplane, they all tried to be of assistance. They cluttered up the frequency with various bits of advice: "Try to keep the wings level," "follow a highway," "fly towards the sun," "turn on the landing light so the airplane can be identified as the one needing help," etc., etc. They kept asking her for her location and to describe her airplane—which only confused her. Many of these airmen offering advice privately did not hold much hope for the rescue of "the lady in distress."

Listening to this babble on the airwaves, a flight instructor, operating out of nearby Statesville Airport, ran to his airplane, a small Cessna 172, to take off and try to do the impossible; find the "voice in the sky" and somehow help her bring the airplane down to a safe landing. He was accompanied by one of his students who wanted to be part of this rescue operation.

In the air, he waited for an opportunity to break through all the transmissions from the other pilots and ground stations which, although well-intentioned, could only be confusing to the panicky recipient. Finally there came a lull, at which point he asked all the other transmitters of advice to please leave the frequency so that he could make radio contact with the endangered aircraft.

He then radioed something along these lines, "Ma'am, what's your name?" At last, a simple and reassuring question. She stated her name. He calmly introduced himself and immediately established a rapport which was comforting and soothing to the lady.

He then asked her to turn the radio knob to another frequency which was not apt to be as busy as the current one. He did the same. Now they could converse without being interfered with by other transmissions. Meantime, he and his student were scanning the horizon still trying to find her airplane.

Locating a small airplane in the sky without any idea which direction to start looking can be an almost impossible task. Nevertheless, he was determined to try. As a start, he followed Interstate 40 (which ran east and west) towards the receding sun and searched the skies—to no avail. There wasn't much time left before it would turn dark. He and his student then scanned both sides of the highway for some distance, without success. After a while, returning to the highway, he headed back towards his own airport and saw—coming towards him—a small Piper Warrior airplane with the landing light on. He held his breath. Could it be? It had to be!

He immediately asked the lady to turn the landing light off, which she did. Now he was positive. He flew in front of her to show that he had her spotted and then maneuvered his airplane just a little behind and to the side of hers so that he could fly in formation and keep an eye on her.

Calmly and quietly, he now started to give her a flying lesson in the air. He explained that he was an experienced flight instructor and that he was going to give her a series of instructions that would enable her to safely control her airplane. The soft tone of his voice and his ability to translate flying terminology into the kind of language that she could understand gave her sufficient confidence in her heaven-sent rescuer to follow his directions without fear or panic.

He told her to move all controls slowly and gently. He complimented her on keeping the airplane on a fairly steady course and had her make shallow turns to the left or right by gently turning the control wheel in those respective directions when necessary to do so. And, after each gradual turn, to level the wings again for straight flight. Little attempt was made to acquaint her with the rudder pedals on the floor of the cockpit which would bring the nose of the aircraft around more quickly—it would only serve to confuse the student in this situation.

She did have some difficulty in locating the throttle, which was a knob on the control panel. She finally identified it when, on his patient urging, she gently pushed the right knob in and out and was able to hear the engine noise increase (when pushed in) and decrease (when pulled out). He explained that the throttle controlled the speed of the engine and therefore the altitude; to go higher, the throttle had to be pushed forward, and to descend, it had to be pulled back. When it came time to land, the throttle would have to be gradually retarded.

As the novice pilot showed signs of being able to control the direction and altitude of the airplane, her instructor gradually guided her towards his home base, Statesville Airport, where a landing would be eventually attempted and where emergency equipment was standing by. This, of course, was transpiring while her husband's body was slumped in the seat next to her and her benefactor's airplane was "riding shotgun" off her wing and just behind her.

It didn't take too long before both airplanes arrived at the outskirts of Statesville Airport. Although she could make out the airport from the air, she had difficulty distinguishing the runway. Therefore, her instructor-guide decided to give her directions to fly over the runway to become familiar with it. They flew over with him radioing quiet commands, such as, "Turn left slowly and gradually" . . . "Now gently straighten your wings" . . . "Push the throttle in slightly to maintain your altitude" . . . "Bring the throttle out just a bit" . . . "Level the wings," and other directives as they passed overhead and then turned around. And the lady complied.

Now, it was time to start the landing approach. In order for the aircraft to safely descend, it was also necessary for her to locate, identify and have some inkling of the functions of other instruments and controls in the cockpit. Calmly, patiently and in an encouraging tone, her instructor proceeded to help her in this most difficult assignment.

It is beyond belief to comprehend the amount of information (which was the barest minimum necessary) concerning the control of the flight of an aircraft that was being fed to a frightened and bereaved person in such a short span of time in circumstances that certainly could not be called a learning atmosphere. Also, in what could be considered (for her) a foreign language. Control wheel, left and right ailerons, rudder pedals, throttle, degree of flaps, revolutions per minute on a tachometer, airspeed indicator, and so on. A scenario set for disaster.

Somehow or other, this woman, with no experience piloting an airplane, was managing to reasonably follow most of these alien instructions until, out of the corner of her eye, she became aware of a movement in the seat beside her. Here she was trying to handle these strange controls at the most difficult stage of flying, the landing approach, when her husband's body began to move—and then started to lean heavily on her. She cried out in terror to the following airplane and explained what was happening as she pushed her husband's inert body back into his seat. Whereupon her instructor decided to delay the landing no longer and immediately gave commands for her to follow in order to land the aircraft.

"Pull the throttle back . . . lower the nose . . . turn the airplane slightly to the left . . . keep the wings level . . . you're dropping too fast, push the throttle in a bit . . . raise the nose . . . level off, level off . . . turn to the right . . . now pull the throttle all the way back—all the way—chop *all* the power . . . line up with the runway . . . you're doing great . . . level your wings . . . get that nose down—more, lower, lower . . . a little more right . . . raise the nose just a little bit . . . level your wings, keep it straight, straight, straight . . . let it settle—easy—let it settle," and finally, "You did it, you did it!" he exulted as the Warrior touched down in the center of the runway.

That joyous exclamation of success was premature. The moment the airplane landed on the runway, her husband's body toppled completely over on her. In addition to the tremendous pressure of following the staccato instructions on landing the airplane, this was too much for her to take. She screamed as his body jolted the controls. The aircraft started to climb back into the air, stalled, fell back to earth, hit hard, ran off the side of the runway onto the shoulder, struck one of the lights with the wing, broke its nose wheel, and pitched forward. The propeller then hit the soft dirt, bent and stopped the engine.

The lady scrambled out of the aircraft onto the ground—unhurt—as the emergency crews and airport personnel rushed to her aid. She asked them to look after her husband (who was beyond assistance) as she walked towards a waiting ambulance on unsteady feet. Sounds of congratulations and relief filled the air from onlookers and radios of ground station personnel and other aircraft.

Two heros emerged from what the National Transportation Safety Board would call "this incident." The flight instructor and, of course, the lady in distress.

The instructor had the ability of taking charge when confusion filled the air in the early stages of the drama. There were innumerable flight personnel and air-to-ground station operators who were willing and anxious to help, but no one was able to organize a coordinated rescue effort until this instructor decided to do something about it.

After isolating their radio conversation, he displayed a marvelous faculty of calming his "new student" and instilling in her the confidence to try to follow his instructions. In addition, perhaps based on his long years of experience in training many different individuals to fly an airplane, he revealed a talent for communicating on a level that was sufficient to have her follow strange instructions without getting too confused. Very few flight instructors would have been able to establish such a bond of communication with someone they had never met before—and in the unlikely circumstances that they found themselves.

The Federal Aviation Administration recognized this unusual ability shown by the flight instructor, who incidentally, was the flight operator of his own airport. On behalf of the National Administrator of the FAA, the Regional Director presented him with a silver medal for Distinguished Service. This award ceremony took place at the home air base of the flying "knight in shining armor."

As for the unscheduled and reluctant pilot-in-training, there had to exist traits of inner strength, determination and fortitude to have her emerge from this experience physically safe and sound. Most people never discover how unique they are until faced with a situation which brings out qualities that they never suspected they had. This lady, who displayed supreme courage in an unbelievable situation, demonstrated that she had those special qualities. And that she is alive and well is irrefutable proof of that fact.

# Chapter 9

# Wind Shear on Landing

We threw in that one story of a small general aviation accident to point out that, despite the regulations requiring regularly scheduled medical examinations for all airmen, there is always the possibility of a pilot suddenly becoming completely incapacitated and leaving his passengers in a precarious position. However, this possible danger is virtually eliminated on the conventional passenger airliner by the regulations requiring at least two flight crew members in the cockpit, both pilots. The wide-body category aircraft calls for a third crew member, a flight engineer. And, in most cases, the flight engineer is also a qualified pilot.

Getting back to the varied causes of accidents that happen to the regularly scheduled commercial passenger airplanes, we now turn to the weather. One form of atmospheric condition that usually makes flight crews apprehensive is known as wind shear. Invisible. To be avoided by pilots, if possible. A simple explanation for wind shear is a condition in which there is a sudden change in wind direction causing both updrafts and downdrafts as well as a change in wind velocity which can adversely affected the airspeed of an aircraft. The problem is that there is no way for pilots to know that wind shear is present until the airplane actually experiences its effect.

Inject this dangerous weather condition when an aircraft is

close to the ground, making its landing approach, and, some-times, anything can happen—and does.

Unfortunately, Eastern Air Lines Flight 66 was in that position at about four o'clock in the afternoon on the 24th day of June, 1975. The three engine 727 jet flight had departed New Orleans, Louisiana, and was now on the final approach, about to land on runway 22L (left) at John F. Kennedy Airport in New York. Including the crew, there were 124 persons on board.

The weather reports for the New York area were far from ideal for landing an aircraft. About 20 minutes prior to the anticipated landing of Eastern Flight 66, the forecast for Kennedy Airport called for thunderstorms, heavy rain showers, with visibilities as low as one-half mile and winds at 30 knots gusting to 50 knots (57 miles per hour). Unfortunately, the flight crew of Eastern 66 did not receive this forecast.

Only shortly before the plane attempted to touch down on run-way 22L, the National Weather Service radar also showed an area of thunderstorm activity centered along the edge of Kennedy Air-port over the approach to runway 22L. According to the National Transportation Safety Board, "There is no evidence that this information was available to either air traffic control agencies or flight crews who were operating in the New York City terminal area."

In addition to not receiving these adverse weather reports, Eastern Flight 66 crew members also were not aware of the fact that only ten minutes earlier, a DC-8 jet, Flying Tiger Line Flight 161, encountered severe wind shifts and came close to disaster when landing on that same runway, 22L. After finally gaining control of the aircraft and completing the landing, the Flying Tiger captain reported to the local controller, "I just highly recommend that you change the runways . . . you have such a tremendous wind shear down near the ground." The controller responded, "We're indicating wind right down the runway [only] at 15 knots when you landed." The captain of Flight 161 replied, "I don't care what you're indicating; I'm just telling you that there's such a wind shear on the final on that runway you should change it to the northwest [another runway]."

The captain of Flying Tiger 161 later testified that during his approach to runway 22L he experienced heavy rain, severe changes of wind direction, turbulence, and downdrafts. He observed wide airspeed fluctuations and at 300 feet had to apply almost maximum thrust to arrest his descent and strive to maintain a safe airspeed. He believed that the wind conditions were so severe that he would not have been able to safely abandon the approach. Therefore, with no alternative, he continued his descent while struggling with the controls of his aircraft, until he successfully landed.

Although the crew members of Eastern Flight 66 were not alerted to the hazardous landing conditions encountered by the Flying Tiger DC-8, they did hear a radio report of another aircraft that actually abandoned its landing attempt after encountering disruptive winds on its final approach. This was another Eastern Air Lines airplane, Flight 902, which was scheduled to land before Flight 66.

Eastern 902 reported to the Kennedy controller, "We had . . . a pretty good shear pulling us to the right and . . . down . . . visibility was nil, nil . . . at 200 feet it was . . . nothing." The Kennedy controller inquired, "Okay, the shear you say pulled you right and down?" Eastern 902 replied, "Yeah . . . the airspeed dropped and our rate of descent was up to 1,500 feet a minute, so we put takeoff power on and we went around [abandoned the landing] at a hundred feet."

Hearing Eastern 902's wind shear report to the controller, the Captain of Eastern 66 said, "You know this is asinine." An unidentified crew member responded, "I wonder if they're covering for themselves [for having abandoned the landing and heading for another airport]."

The controller asked Eastern 66 if they had heard Eastern 902's report. Eastern 66 replied, "Affirmative." The controller then cleared the flight for an instrument landing approach to runway 22L (despite the fact that both Eastern Flight 902 and Flying Tiger 161 had reported experiencing dangerous wind shear conditions with almost disastrous consequences on attempting to land on that runway). Eastern 66 acknowledged the clearance and prepared for the landing.

At this stage, we can only conjecture why, after hearing this negative report, Eastern 66 decided to continue with its approach to runway 22L. Having flown in from New Orleans, the flight crew probably did not relish the thought of abandoning the scheduled landing at J.F. Kennedy and being diverted to another airport (La Guardia, Newark or even, depending on weather, Boston or Philadelphia). After all, some of the crew probably had their cars parked at Kennedy, or had made arrangements to be picked up by others. Plus the fact that there had been so many instances in the past in which, despite severe weather reports, the aircraft experienced little difficulty with the landing. Sometimes this leads to a form of compulsive behavior in pilots that is known in aviation circles as "get-there-itis."

The First Officer of Eastern 66, who was flying the aircraft, called for completion of the final checklist. While the checklist items were being reviewed, the captain said, "I have the radar on standby in case I need it [in the event he comes to a decision to abort the landing]." Obviously, the captain of Flight 66 was a little concerned about landing on runway 22L.

Shortly before Eastern 66 was due to start its landing approach, the controller sought confirmation on the existence of severe wind conditions from the airplane that abandoned its landing approach, and asked Eastern 902, "Would you classify that as severe wind . . . shear?" Eastern 902 responded, "Affirmative."

Hearing this, the first officer of Eastern 66 said, "Gonna keep a healthy margin on this one [extra airspeed as a margin of safety to prevent too rapid a descent or a possible stall]." An unidentified crew member said, "I . . . would suggest that you do." The First Officer responded, "In case he's right." (The crew was apprehensive but determined to go ahead.)

The controller then cleared Eastern 66 to land and the captain, acknowledging, asked, "Got any report on braking action?" The local controller replied, "No, none, approach end of runway is wet . . . I'd say about the first half is wet—we've had no adverse reports."

It was now a little after four PM. Flight 66 was centered on the glideslope (the radio beam aircraft use to descend to the runway in instrument-landing conditions) when the flight engineer called out the altitude, "500 feet." The sound of heavy rain could be heard as the aircraft descended below 500 feet, and the windshield wipers were switched to high speed.

The Captain said, "Stay on the gauges." The First Officer responded, "Oh, yes, I'm right with it."

At this point, another aircraft asked the tower controller, "Everyone else . . . having a good ride through?" The tower responded, "The only adverse reports we've had about the approach is a wind shear on short final."

Getting back to the cockpit conversation, the Captain announced, "I have [see] approach lights," and the First Officer, flying the airplane, responded, "Okay." Again the Captain cautioned, "Stay on the gauges," to which the First Officer replied, "I'm with it."

When the aircraft got down to 400 feet, the rate of descent suddenly increased dramatically from 675 to 1,500 per minute (a much faster descent than normally called for at this stage of the landing approach). The airplane now rapidly began to deviate below the glideslope (aircraft should stay centered on the glidepath, certainly not below it). A few seconds later, due to a sudden shift in wind direction, the airspeed decreased, hastening the descent. The airplane was getting tougher and tougher to control.

Eastern 66, fighting to maintain its heading towards the runway, continued to deviate dangerously below the glideslope. As the aircraft descended to only 150 feet above the runway, the Captain, peering out the cockpit window, called out, "Runway in sight." The First Officer, who was still at the controls, affirmed, "I got it." However, a second later, there was an unintelligible exclamation recorded on the cockpit voice recorder followed by the First Officer urgently commanding, "Takeoff thrust!" (He either noticed that the airliner was at a dangerous position in relation to the runway or the aircraft was suddenly shoved downward by a severe gust of wind.) A desperate attempt was made to apply full power to all three engines to arrest the descent and somehow propel the aircraft upwards.

Too late.

Eastern 66, having safely flown over six approach light towers, struck the top of the Number 7 light tower with a devastating impact. (The approach light system at Kennedy Airport, to guide the pilot to the runway, consists of a series of high-intensity lights mounted on top of steel towers standing up to 30 feet high, in line with the runway.) The collision with the light tower took place less than a half-mile short of runway 22L.

After its impact with tower number 7, the aircraft, still in the air, continued flying and struck towers number 8 and 9 (the towers are located 100 feet apart). At this point, the left wing shattered and the outer section of the wing was severed from the rest of the airplane. Eastern 66 then rolled into a steep left bank and caromed into the terrain at high speed. Now on the ground, it tore past, and just missed, towers 10, 11 and 12. However, continuing its journey of horror, its fuselage now struck, in turn, towers 13, 14, 15, 16 and 17—leaving a trail of victims and debris behind at each tower site. What was left of the airliner (two separate sections of the fuselage and one section of a wing), scraped, bumped, skidded and slid along the ground, out of the airport boundary, up to and across a street in Queens, New York, where it finally came to rest. The remainder of the airplane, starting from the first tower struck, had disintegrated into thousands of bits of metal, seats, wire, cushions, pipes, fabric, plastic, knobs, dials, rivets, nuts and bolts.

Fire had broken out when the left wing tore off, spilling fuel as the aircraft skidded through the approach light towers. There were numerous ignition sources—hot engine components, electrical wiring in the aircraft, the approach lights, and the street light system. Destruction of the fuselage caused more fuel to be released, and the fire continued to burn after what was left of the airplane came to rest.

Along the wreckage path were the dead, injured, luggage, articles of clothing, jewelry, money in denominations up to $100, cosmetics and endless personal items. Bodies were on both sides of a wide road called Rockaway Boulevard and some even on a nearby garbage dump.

All four members of the flight crew and two of the four flight attendants aboard Eastern 66 perished in the crash. The two surviving flight attendants were seated in the rear of the passenger cabin.

In addition, 12 passengers of the 116 on board were also among the survivors. They were seated in the rear section of the passenger cabin which, together with the tail of the aircraft, remained relatively intact.

(Theories have been advanced that in the event of a crash, if there are survivors, they are more apt to have been seated in the rear of the aircraft. This was true of the DC-10 accident in Mexico City which was covered in Chapter 4 and will also turn out to be true in subsequent chapters that you will be reading.)

Viewing the extent of the wreckage, it is hard to believe that any of the occupants escaped death. The accident generally was not survivable because of the destruction of most of the aircraft's fuselage. The cockpit and forward flight attendants' seats along with the passengers' seats were torn from their supporting structures. These seats were mangled, twisted, and scattered along the area that the aircraft traveled. Only the aft flight attendants' seats remained in place.

Rescue and livesaving efforts were both heroic and speedy. It was truly a dedicated team effort by different teams of professionals.

Firemen arrived almost immediately after the accident and extinguished the raging fire within minutes. Their rapid response prevented fatal burns to the passengers who ultimately survived; some of them were found lying in pools of fuel and fire-extinguisher foam.

The firemen also rescued a teenager trapped in a seat between two dead passengers and rushed him in their firetruck, along with five other survivors, to Jamaica Hospital before the ambulances even arrived at the scene. The ambulances got there moments later and sped the remaining eight survivors to the same hospital.

Doctors were there waiting for the injured while other medical personnel were dashing from their homes after receiving news bulletins of the disaster. Everyone participated (specialists, staff physicians, residents, interns, nurses, technicians, therapists, orderlies, etc.) in trying to save the survivors with the realization that some were beyond help. Their task was complicated by the fact that the extensive burns had to be dealt with before internal injuries could be treated or broken bones set. Needless to say, in the very early stages, there was a good deal of rushing about and confusion before an organized program emerged.

Physicians at Jamaica Hospital were assisted by doctors from other hospitals, including Jacobi Hospital located in the Bronx, which is world-renowned for its burn center. Four of these patients were then transferred to Jacobi via medical van, receiving treatment enroute, where they could be treated by burn specialists employing the latest sophisticated burn equipment available at that hospital.

Every possible care was taken to give the injured (all of whom were burned in addition to suffering from either internal injuries or broken bones, or both) a chance to survive. Cost was disregarded. Hospital personnel dealing with the injured were ordered to wear masks, gloves, caps and gowns to help prevent dangerous infections to the victims. The number of medical personnel assigned to each of the patients was extraordinary. The chief of surgery at Jamaica Hospital said, "We have three or four doctors as well as two or three nurses caring for each patient, plus lots of support personnel."

Despite the heroic efforts and superior medical care, two of the 14 injured survivors died shortly after they arrived at the hospital. Two other passengers died within five days of the accident and one passenger died nine days after the crash. However, nine passengers, along with the two surviving flight attendants, were on the road to recovery.

The crash attracted hordes of curiosity seekers. Cars filled the nearby streets until the police came to clear the roads so that emergency vehicles could get through.

According to the police, two people were arrested at the accident site for impersonating emergency workers and police officers. Said a police spokesman, "They were either curiosity seekers or thieves." He added, "We don't know which. One guy had a doctor's bag and white coat and another took a coat, hat and boots off a fire truck."

Autopsies were performed on all of the people who died in the crash. The deputy chief medical examiner of New York City stated that the victims were subjected to forces that were "tremendous, more in this crash than the others I have seen [victims of other airline crashes]."

He said that most of the victims died immediately from the impact. Heads and arms were propelled against the seats in front of them with such force that they received multiple fractures of wrists and forearms. Many of them had their skulls burst open and others died from internal bleeding caused by the sudden tremendous force against the fastened seat belts.

As a result of these autopsy findings, this chief physician said, "If these people were sitting backwards, I think a lot of these injuries might have been diminished because, in this particular crash, they didn't die so much of smoke inhalation, suffocation, burns, or weren't impaled on tree branches after having been ejected from the plane. They died within seconds of impact injuries."

The deputy chief medical examiner continued, "Many things we've learned about air safety we've learned from accidents like this." He said, "When I travel, I travel in the last seat in the coach. It's amazing, in a crash, how many people in the front of a plane, in the first class section, get 'wiped out.'"

The National Transportation Safety Board was notified of the accident and immediately went into action. As part of the investigation, a public hearing was held. Parties to the hearing were: The Federal Aviation Administration, Air Line Pilots Association, Professional Air Traffic Controllers Organization, Eastern Air Lines, Inc., and the National Weather Service.

Four different controllers, with different responsibilities for directing traffic at Kennedy during the accident time period, testified at the Safety Board's public hearing.

It is interesting that all four controllers stated that they were aware of thunderstorm conditions in or near the approach path of landing on runway 22L, and yet not one of the four made the recommendation (as the Flying Tiger Line pilot did) to change the runway. All of them were very busy with their duties and a change of runways would have called for a change in the approach pattern, with further delays in landing to all flights terminating at Kennedy.

In addition to the testimony of the four controllers, a number of airline pilots stated that when they conduct instrument approaches to airports affected by weather hazards they rely substantially on the experiences of pilots who precede them when they decide whether to make the approach themselves or to choose a different course of action. Unfortunately, Flight 66 did not receive the report of the landing experience of Flying Tiger 161.

What also came out of the hearing was that during the year preceding the accident, Eastern issued a number of bulletins on low-level wind shear associated with thunderstorms. The bulletins implied that higher approach speeds should be used when shear is anticipated.

The flight crew of Eastern 66 had added 10 to 15 knots to their normal approach speed, which the Safety Board believed was reasonable. However, even with this airspeed margin, the pilot should have immediately recognized the aircraft's descent below the glideslope and made rapid changes to prevent the airplane from landing short of the runway.

The Safety Board explored the reasons why approach operations to runway 22L were continued, particularly after both pilots and controllers had been warned that severe wind shear conditions existed along the final approach to the runway. Especially since the thunderstorm astride the course to runway 22L was obvious and since there was a relatively clear approach path to another runway, 31L.

The Board eventually concluded that a runway change did not take place for several reasons:

1.  Runway 31L had already been in use for six hours prior to runway 22L's being made the active runway. The noise abatement program (which called for rotating runways after a number of hours of use to spare the local community from too long an exposure to noise pollution) now favored the use of runway 22L, especially since that runway was most nearly aligned to the wind.

2.  A runway change, at the time, would have increased the already heavy workload of the controllers.

3.  The air traffic system in a high density terminal area tends to resist changes that disrupt or further delay the orderly flow of traffic. Delays have a compounding effect. Consequently, controllers and pilots tend to keep the traffic moving because delays involve the consumption of fuel as well as missed connections

with other flights.

The Safety Board also investigated the performance of the 727 jet aircraft and the pilot under the adverse wind conditions encountered. With the cooperation of the Boeing Company, the manufacturer of the aircraft, a series of tests was conducted which was performed on a flight simulator programmed with the dynamic winds and the flight characteristics of Eastern Air Lines Flight 66. The objectives of the simulator tests were to examine the flight conditions which probably confronted the flight crew of Eastern 66, and to observe the difficulties that a pilot has in recognizing the development of an unsafe condition and in responding with appropriate corrective action.

Fourteen pilots participated in the tests. Each pilot flew several approaches through one or more of the wind models. The pilots were told to attempt to maintain an airspeed which was 10 to 15 knots above the normal speed on approach. They were given the option of attempting to land or executing a missed-approach, but, in any event, they were to try to avoid landing short of the runway threshold (which is what happened to Flight 66).

Fifty-four approaches were flown. On 31 of them, the pilots, prior to touchdown, abandoned the attempt to land and executed a missed-approach. On 18 of the approaches, the simulator descended to an altitude that indicated a collision with the approach lights. Only five approaches of the 54 were successful in placing the simulator on the runway in a position from which a safe landing could be made.

Following the simulator tests, comments were solicited from the 14 pilots participating in the tests and 10 responded. Eight of the 10 pilots believed that they probably would have crashed during actual flight in comparable conditions encountered by Flight 66.

After completing its investigation, the Safety Board issued a statement saying that although the flight crew should have relied more upon their flight instruments to correct the high descent rate, the adverse winds might have been too severe for a successful landing in any event. And, "Contributing to the accident was the continued use of runway 22L when it should have become evident to both air traffic control personnel and the flight crew that a severe weather hazard existed along the approach path."

Later, a considerable number of lawsuits were filed against Eastern Air Lines. The U.S. government was also a defendant because the suits contended that the air traffic controllers should not have permitted the airplane to land.

By now, almost all of the suits have been settled for an estimated $30 million. The most recent settlement to the family of one of the victims amounted to $4 million.

A postscript of interest: Two major ingredients of the tragic ending of Flight 66 were 1) the inability of the controllers to detect wind shear, and 2) the collision of the airplane with the "unforgiving" steel structure of the approach light towers. One can then say that the only good to emerge from this particular air disaster is that the authorities have finally taken steps to eliminate these two deadly situations.

Let's consider the problem of the steel light towers first. The need for frangible towers that will "give" when struck by an aircraft had been recognized even before the crash of Flight 66. An order calling for the revising of construction of all approach light towers was issued three months before the accident by the Federal Aviation Administration. This order provided that frangible structures would be used for all future approach light installations. In addition, a retrofit program would be adopted, as funds were made available, to change all present nonfrangible approach light towers to ones that would give an aircraft an opportunity to survive contact with the towers.

The type of towers now being installed are designed to fracture and collapse at impact speeds of as low as 80 knots (92 miles an hour) and probably would give way at speeds well below that depending on the type of aircraft involved. (A 727 Boeing jet passenger aircraft, at its normal landing speed, would be able to brush the tower aside and continue its glide path safely to touchdown.)

After the crash, Kennedy Airport's runway 22L was moved up on the priority list for the change. The replacement of the nonfrangible approach light towers by the more flexible frangible towers for runway 22L was started in 1979 and completed in early 1980.

As for the wind shear problem, three years after this accident at Kennedy, a program for installing wind shear detectors at all major airports in the United States was initiated. This equipment in quite effective in detecting gust fronts, wind shifts, and strong gusty winds at a level as low as 15 feet above the runway. These detectors are monitored by controllers in the tower.

To date, more than 50 major airports have installed this important equipment. Additional funds are currently being sought to outfit the towers of additional airports.

Speaking of the presently installed wind shear equipment, a spokesman for the FAA said, "We haven't lost an aircraft through wind shear at these airports since installation."

Chapter 10

# Wind Shear on Take-Off

Those late reassuring words by the FAA official were spoken too soon. Unfortunately, we have another wind shear inside story to write about. This accident, although different from the one you just read, is similar in that it also had an unhappy ending.

The Kennedy disaster involved an airplane coming in for a landing—with reduced engine power. The one you are about to read happened to a passenger jet airplane just after takeoff—using maximum thrust.

Both were operating at opposite ends of the power curve. Minimum power or maximum power; it didn't matter. Both succumbed to the wind shear peril. Let's follow the flight of the more recent tragedy.

On July 9, 1982, a Pan American World Airways Boeing 727 three-engine jet, Flight #759, departed from New Orleans International Airport, Kenner, Louisiana, heading for Las Vegas, Nevada.

At the time of takeoff, 4:08 PM, there were showers over the airport and along the airplane's intended takeoff path. The winds were gusty, variable and swirling, and although the Captain had been cautioned several times that low level wind shear alerts were being registered on the recently-installed wind shear detection equipment, his decision to take off, according to the later issued

Safety Board report, "was reasonable in light of the information that was available to him."

The Safety Board also noted that the Captain advised the flight engineer to turn off the air-conditioning packs for the takeoff (which would increase the thrust capability of the engines) and also directed the First Officer, who would be at the controls during takeoff, to "let your airspeed build up on takeoff [extra speed before lifting off]."

According to witnesses, Pan Am Clipper 759's liftoff and initial climb in a wing-level position appeared to be normal until the aircraft reached an altitude of about 100 to 150 feet above the ground, and then the airplane began to descend. Flight 759, with no warning, had run into a microburst—a small-scale weather phenomenon, short in duration, involving high velocity downward-moving air spreading out rapidly in all directions as it nears the ground. This burst of air literally pushed the three-engine jet aircraft towards the ground and, in less than six seconds, the altitude was down to 50 feet.

The captain called out, "Come on back [up], you're sinking, Don . . . come on back." This was immediately followed by the actuation of the aircraft's Ground Proximity Warning System, "Whoop . . . whoop . . . pull up . . . whoop . . ." The conclusion of the Safety Board afterwards was that the First Officer "had probably reacted and was applying corrective action," and that his correction of the aircraft's settling towards the ground "equalled the response which could be expected under the prevailing conditions."

However, although it appeared that the descent was arrested, the aircraft struck three trees at the 50-foot level less than a half-mile beyond the departure end of the runway. Leaving some parts of the airplane behind, the airliner then struck a second group of trees with its left wing and started to bank towards the left. Continuing its roll onto its left side, it finally struck the ground with its left wingtip and crashed into a residential area less than one mile from the end of the runway.

All of the 145 occupants of the aircraft, including the seven crew members, were killed by the impact, immediate explosion and subsequent fire. Eight persons on the ground also perished.

Six homes were completely destroyed and five others were damaged substantially over the three-block area devastated by the airplane falling out of the sky.

The National Transportation Safety Board, after its usual investigation, concluded that "the probable cause of the accident was the airplane's encounter during the liftoff and initial climb phase of flight with a microbust-induced wind shear which imposed a downdraft and a decreasing headwind, the effects of which the pilot would have difficulty recognizing and reacting to in time for the airplane's descent to be arrested before its impact with the trees.

"Contributing to the accident was the limited capability of current ground-based low-level wind shear detection technology to provide definitive guidance for controllers and pilots for use in avoiding low-level wind shear encounters."

As an aftermath of the tragedy, there are three items of human interest worth mentioning:

1. Seven members of one family from Louisiana were passengers on the airplane. They were on their way to a funeral in Las Vegas for another family member.

2. Five women friends (four nurses and a pediatrician) who had planned a fun trip to Las Vegas for more than five years, were bumped from the Pan Am flight because it was oversold. They were able to be placed on a plane from another airline but their luggage could not be retrieved and remained aboard the ill-fated Pan Am airplane.

The five friends were almost in shock when they later learned about their narrow escape. In Las Vegas, Pan Am presented each of them with a "survival kit" (which included a toothbrush), $300 to buy clothes, and also reimbursed them for the cost of the ticket to Las Vegas. However, the luck of the "Fortunate Five" did not carry over to the casinos—they lost. When asked how they felt about flying home, one responded for all of them, "Terrified."

3. Several hours after the Pan Am crash, a rescue worker was searching through the rubble in one of the suburban houses decimated by the aircraft, without much hope of finding any survivors. Suddenly, out of the corner of his eye, he noticed a

movement—a mattress lifting up slightly from the ground. The mattress was completely lifted off and underneath was a 16-month-old little girl. She was on her stomach with her hands spread out in front as if trying to lift the load off her back and stand up. She was wide awake, looked up at this strange man and began to cry. Said her rescuer, "It broke my heart."

The baby, wearing only a diaper, was rushed to the hospital where she was reported to be in excellent condition with minor burns on her left hand and both feet. The little tot was the only survivor in her house; her four-year-old sister and 26-year-old mother were among those who were killed. Her father was at work at the time.

This accident at New Orleans pointedly demonstrated the fact that the ground installation of wind shear equipment at airports, although a major step in detecting this dangerous condition, was not enough. Thus, when one airline announced that it was planning to install airborne wind shear detection devices in all of its aircraft, the news was greeted with great interest.

The airborne system is designed to tell a pilot who has not received a warning from the ground, not only that he is entering a wind shear, but also the best maneuver to fly out of it. Said a spokesman for the airborne device, "Our system puts the solution in the cockpit, where it must be, to protect the aircraft when a hazardous encounter occurs."

The Federal Aviation Administration is in favor of these detection devices in aircraft. In 1987, the FAA announced that, in addition to making it mandatory for airline pilots to undergo a training program in coping with wind shear, it will be proposing that all airlines install airborne equipment.

# Chapter 11

# Both Engines Flame Out in Thunderstorm

Enough of wind shear. It is time to turn to a different type of weather-induced accident.

This one has to do with thunderstorms, another atmospheric condition that pilots would rather not encounter. The effect of these storms on the engines of the unlucky aircraft you will be reading about was something, evidently, that the airline industry had not contemplated even as a remote possibility.

In the first 20 years of jet airline travel, there had been no recorded instance of any passenger airplane *ever* having to make an emergency landing with *all* of its engines not operating. Therefore, it would appear to be a sheer waste of time and money for the airlines to require this type of training and practice for their jet pilots. Right? Wrong.

The case we are about to read about is a first time for this type of emergency. Unfortunately, it was a costly initiation, especially in terms of human lives. And, as fate would have it, Southern Airways Flight 242 was chosen for this dreadful distinction.

It was raining quite heavily and was very windy on April 4, 1977, when a DC-9 twin-engine jet aircraft took off from Huntsville, Alabama. The time was 3:54 PM. The airplane's destination—the Atlanta Airport in Georgia.

It was not the kind of weather in which pilots enjoy flying. However, the Captain and his First Officer were not concerned. They were experienced professionals. They had flown in bad weather before and the aircraft did have radar equipment to help guide them through the rougher air. As for the 81 passengers (and two flight attendants) on board Flight 242, like most airline travelers, they had the utmost confidence in the capability of the two pilots to deliver them safely to their destination.

Since the estimated time for that short run to Atlanta was 25 minutes, the flight crew requested and received permission to fly at an altitude of only 17,000 feet, hoping to avoid most of the turbulent weather expected en route. Shortly after, the controller told Flight 242 that his radarscope was showing heavy precipitation and asked the pilot what his flight conditions were at the time. He received the reply that, so far, while climbing towards 17,000 feet, the aircraft was experiencing only light turbulence. So far, so good. The controller acknowledged Flight 242's report and told the flight to contact the next controller at Memphis Center.

The Memphis Center controller's report was not good. He advised the flight that extremely severe weather conditions were prevalent in the area and that a SIGMET was currently in effect. (SIGMET is a weather advisory concerning weather significant to the safety of all aircraft. It includes tornadoes, lines of thunderstorms, large hail, severe turbulence and dust or sandstones.) The controller suggested that Flight 242 monitor SIGMET radio broadcasts and then told the crew to establish communications with Atlanta Center. As the Captain acknowledged this transmission, the aircraft ran into heavy turbulence. The Captain, fighting to control the lurching, bucking aircraft, was heard to say, "Here we go . . . hold 'em cowboy."

Warily keeping the heavily-jiggling aircraft under control, the crew contacted and informed Atlanta Center that Flight 242 was climbing out of 11,000 feet heading for the assigned altitude of 17,000 feet. As it climbed, the pilots had their hands full. The airplane was encountering heavy turbulence, heavy rain and pellets of hail. Visibility from the cockpit was nil and the aircraft was jolting, quivering and bouncing from side to side as well as up and down. Unpleasant for the pilots, and, of course, unnerving for the passengers.

As a safety measure, to reduce the strain and load on the wings due to the turbulent gusts, Flight 242 reported to Atlanta Center that they were reducing the speed, as the sounds of rain and hail were recorded in the background. Then, trying to locate the thunderstorms shown on the aircraft's radar, the pilots decided to go through a hole depicted on the screen as one which would subject the aircraft to the least amount of turbulence. As they steered through this "opening," the sounds of heavy hail and rain continued to be recorded until, suddenly, a bolt of lightning glanced off the airplane, affecting the electrical system.

Atlanta Center made four transmissions to Flight 242 during the next 36 seconds; without a reply. When the radios returned to functioning, the co-pilot was heard to say, "Got it, got it back."

The events that have occurred up to now, are turning out to be ominous: taking off in the face of severe weather forecasts, running into heavy rain, hail and turbulence, temporarily losing control of the aircraft, and now, a bolt of lightning followed by an interruption of radio communication. One unfavorable incident triggering another.

Atlanta Center, resuming communications, asked Flight 242 to maintain 15,000 feet. However, the response the controller received was that the aircraft was having difficulties. That it was 14,000 feet and would try to get back up to 15,000, but now there was another problem—the windshield had been smashed in that last bout with the turbulence.

If that wasn't enough to upset the crew, a few moments later, Flight 242 was confronted with a more serious situation; the left engine cut out completely. As they relayed this information to the Atlanta Center, within 30 seconds, the impossible happened—the other engine stopped operating.

When the pilot radioed this last bit of chilling news, the controller, having some difficulty comprehending this turn of events, requested, "Say again." Flight 242 replied, "Standby, we lost *both* engines."

(Later, an official from the Federal Aviation Administration, when asked a question about the flameout of Flight 242's jet engines, said, "An engine can fail due to turbulence or heavy intake of rain and hail. And the plane was going through rain with hail at the time."

(He went on to state that a jet engine depends on a "steady stream of air coming in. And if you get into turbulence, this steady flow can be interrupted.")

At 14,000 feet, with both engines out and no thrust available, Flight 242 had no choice but to try to glide to an area, hopefully free of obstructions, where the aircraft could make an emergency landing. The pilot lowered the nose slightly to gain maximum gliding distance and asked Atlanta Center to direct him to a clear area.

Atlanta Center, now fully aware of the extreme emergency, advised the flight crew to contact the Atlanta Approach controller, who could direct them straight into Dobbins Airfield, located about 20 miles from their present position. The co-pilot, telling the Captain that he was familiar with Dobbins, took over the controls. He then informed the Atlanta Approach controller that his engines were not working, that he was down to 7,000 feet and asked for a vector (heading) to Dobbins.

While being directed to Dobbins, the pilots frantically kept trying to get at least one engine going, but were unsuccessful. With 17 miles yet to go, they were now only 4,600 feet above the ground. Although the controller, at this point, informed them that there was another airport only 10 miles away, the crew realized that they would run out of airspace before the aircraft could glide even that distance.

For the next 90 seconds, the two pilots looked for a place to set the aircraft down. As the airplane, in its glide, descended lower and lower, they debated whether to choose a field or a roadway. Finally, forced to make a decision, they picked out a highway that, although not straight, appeared, from the air, to be free from automobile traffic. (Unfortunately, it didn't quite turn out to be that way.)

(Afterwards, a pilot from another airline, monitoring the radio conversation of Flight 242, had the highest praise for the pilot. "He did everything he could do to get that plane down under the most difficult circumstances. He was trying to restart his engines, talking to the controllers, steering and trying to keep his glide."

The other pilot pointed out the fact that without power, a DC-9 airliner falls one foot for every 10 feet it travels. Since the aircraft was gliding at about 200 miles an hour and falling at 20 miles an hour, he said, "At those speeds [at a low altitude], you don't have too much visibility. Highway 92 was the only place left on which to put the aircraft.")

At about 4:18 PM, with the Captain cautioning his co-pilot, "Don't stall it" Flight 242's last transmission was recorded: "We're putting it on the highway, we're down to nothing."

With the First Officer trying to coax the DC-9 airliner gently down on State Highway 92, the outer section of the left wing first struck two trees just outside the community of New Hope, Georgia. Still airborne and still under directional control, the airplane began to strike trees and utility poles on both sides of the road (State Highway 92 was not wide enough to accommodate the wingspan of the large aircraft) until, finally, the main gear touched down on the highway. Up to this point, the occupants of the airplane were still unharmed because the fuselage was intact and only the wings had been impacted. However, their good fortune was not to last.

Almost simultaneously with the touchdown, the tip of the left wing struck an embankment. Now the aircraft veered uncontrollably to the left and off the highway into the New Hope Community. With the pilots helpless to control it, the aircraft became a travelling bomb. Spitting fuel and bursting into flames, it struck automobiles, trucks, fences, trees, houses, shrubs, lawns, a combination grocery store-gasoline station with its gas pumps, highway signs, poles, powerlines and everything and anything that got in its way. The debris and chaos left behind was considerable; the wreckage area was 1,900 feet long and 295 feet wide.

People who were driving on the highway at the time, saw the huge airplane suddenly appear from nowhere descending upon them. Said one witness, "It looked like he was gonna come crashing through my windshield." He estimated the plane was no more than a few feet over his car. He was lucky to escape unscathed.

Another automobile was not that fortunate. Out on the road, with no warning of what was about to happen, seven young occupants (the oldest 23-years-old), all related to each other, were killed when their car was struck by the airliner. The victims consisted of two sisters with their three children, and their sister-in-law and her youngster. A terrible tragedy to one family who happened to be in the wrong place at the wrong time.

Before the crash itself, passengers went through a grueling and terrifying experience. They became apprehensive about their flight when they began receiving emergency instructions from the flight attendants.

According to the Cockpit Voice Recorder, shortly after the aircraft entered the heavy hail and rain, the aft flight attendant announced on the cabin address system that the passengers should keep their seatbelts securely fastened and instructed them about what to do in the event of an emergency landing.

Several minutes later, when certain that both engines were not operating, the flight attendants began to brief the passengers on evacuation procedures: they demonstrated how to open the exits, and how to assume the brace position on command. Additionally, they instructed the passengers to remove sharp objects from clothing, to check that luggage was stowed securely, and to remove their shoes to prevent damage to the evacuation slides during evacuation.

After the briefings, the forward flight attendant opened the cockpit door to tell the flight crew that the passengers were prepared for an emergency landing, but the First Officer snapped at her and ordered her to return to the cabin and sit down. She called the aft attendant on the interphone and told her about the situation in the cockpit and they discussed their preparations for an emergency landing and evacuation.

Shortly thereafter, the aft flight attendant saw trees outside the cabin window and yelled to the passengers, "Grab your ankles!" The forward flight attendant repeated the command, and the passengers responded as instructed. There were no signals from the flight crew that landing was imminent. According to the stewardesses, they received no information from the flight crew about what had happened after the aircraft entered the heavy hail, or how the flight crew planned to land the aircraft.

During the crash, many of the passengers were ejected from

the aircraft, some still strapped in their seats. However, it is worth mentioning that two survivors, who were seated in the last row of the airplane, reported their seats remained firmly in place.

One of them, lying in the hospital, recounted an almost insignificant occurrence that probably saved his life.

"I usually sit in the middle next to a window," he said. This time, he walked past a number of empty seats to select one in the back row. That decision enabled him to become one of the survivors.

"I called my wife last night after I was admitted here," he said. "She first heard that everybody on board had been killed. Then she heard that some had survived. She told me she didn't think I was dead because I was too mean."

Volunteer firemen who witnessed the crash from a nearby fire station in New Hope responded immediately to the crash scene. These firemen were assisted by others from the Hiram and Union volunteer fire departments as well as the regular firemen from the Cob County Fire Department. The fires were extinguished in about 30 minutes.

Residents of the small community whose homes were located near the crash site were shocked at the condition of the survivors. One woman, who lived close by, said, "About 20 people came rolling out into the grass. Some of them were on fire and came running to my house with their skin burning off.

"They were begging for water. They were hurting. I couldn't get water because the electricity was out and my water wouldn't pump. They went to my refrigerator and got ice out of the ice-maker."

Her husband, returning from work, related the reaction of his three children when they initially saw the burned survivors. "They were terrified," he said. "My two-year-old told me later, 'Daddy, there were monsters.'"

Unfortunately, the survivors were far outnumbered by those who perished in the unscheduled landing.

When the authorities were finally positive of the fate of all of the occupants of Flight 242, as well as the individuals on the ground physically affected by the accident, an official injury list was released. It showed as follows:

| Injuries | Crew | Passengers | Individuals on the Ground | Total |
|----------|------|------------|--------------------------|-------|
| Fatal    | 2    | 60         | 8                        | 70    |
| Serious  | 1    | 21*        | 1*                       | 23    |
| Minor    | 1    | 0          | 0                        | 1     |

*Two persons died about one month after the accident, but were not listed as fatalities because, according to a flight regulation, a fatal injury is defined as one which results in death within seven days after the accident.

Following are some miscellaneous statements made by or to the Safety Board investigating the accident:

1. Standard operating procedures dictate that a Captain take control of the aircraft in an emergency situation. It could not be determined why he did not take over control in the final stages of the emergency landing. His total flying experience was far superior to that of the First Officer. It can be theorized that his greater familiarity with the DC-9 and its systems made it logical that he devote his attention to attempts to restart the engines. The Captain may also have considered the First Officer's familiarity with Dobbins and its approaches a reason to let the co-pilot handle the aircraft.

2. According to a post-accident analysis by the National Weather Service, the storm system that moved across Alabama and Georgia that afternoon was one of the most severe systems in the United States in the past three years. Also, it was one of the fastest moving systems on record. About 20 tornadoes and 30 severe thunderstorms were included in the system.

3. One of the passengers, a commercially licensed pilot, testified that the flight was routine until the aircraft encountered severe turbulence followed by very heavy precipitation, a lightning strike on the left wingtip, and hail. The hail increased in intensity and size; then one engine quit, and the other shortly thereafter.

4. The aircraft manufacturer provided information on the glide ratio of a DC-9 aircraft with engines inoperative. Under the atmospheric conditions that existed on the day of the accident, the

aircraft could glide about 34 miles in wings-level flight while descending from 14,000 feet—which is just about what Flight 242 was able to do that day.

5.   The feet of a number of the survivors were cut and burned because they had no shoes for protection. In accordance with standard evacuation procedures, the flight attendants had briefed the passengers to remove their shoes to prevent damage to evacuation slides. Because of the lack of information from the flight crew, the flight attendants had no way of knowing the circumstances associated with the landing and, therefore, had no reason to deviate from standard procedures. Although the flight crew was preoccupied with trying to restart the engines and with selecting suitable landing areas, the Safety Board concluded that a few words to the flight attendants about the type of landing expected, might have enabled the attendants to better prepare the passengers. Had pillows and blankets been distributed and had shoes been worn, some of the passengers' injuries would have been less severe and more passengers would have been able to escape from the wreckage.

To truly capture the sequence of events leading up to the accident, one should listen to the conversation of the crewmembers as they performed their duties during that tragic flight.

### TIME: 3:56 PM Eastern Standard Time

Huntsville Controller:     Southern 242, I'm painting a line of weather [showing thunderstorms on radar screen] which appears to be moderate to possibly heavy precipitation starting about five miles ahead.

Captain:     Okay, we're in the rain right now, it doesn't look much heavier than what we're in, does it?

Huntsville Controller:     It's not a solid mass, it appears to be a little bit heavier than what you're in right now.

| | |
|---|---|
| Captain: | Okay, thank you. |
| Huntsville Controller: | Southern 242, you're in what appears to be about the heaviest part of it right now, what are your flight conditions? |
| Captain: | We're getting a little light turbulence now and I'd say moderate rain. |
| Huntsville Controller: | Okay, it won't get any worse than that, and contact Memphis Center on [frequency] 120.8. |
| Captain: | [1]20.8, good day, and thank you much. |
| Captain: | Memphis Center, Southern 242 is with you climbing to one seven thousand [feet]. |
| Memphis Center: | Southern 242, Memphis Center, roger. |
| Captain: (Intra-cockpit) | As long as it doesn't get any heavier, we'll be all right. |
| Co-Pilot: (Intra-cockpit) | Yeah, this is good. |
| Memphis Center: | Southern 242, contact Atlanta Center [on frequency] 134.05. |
| Captain: | One thirty-four zero five, good day. |
| Memphis Center: | Good day. |
| Captain: (Intra-cockpit) | Here we go, hold 'em cowboy [evidently heavy turbulence]. |
| Captain: | Atlanta Center, Southern 242, we're [climbing] out of eleven for seventeen [thousand feet]. |
| Atlanta Center: | Southern 242, Atlanta Center, roger. |
| Cockpit radio: (intra-cockpit) | (Sound of hail and rain). |
| Public Address System in aircraft with stewardess B in rear cabin speaking: | Keep your seatbelts on and securely fastened, there's nothing to be alarmed about, relax we should be out of it shortly. |
| Atlanta Center: | Southern 242, what's your speed now? |
| Atlanta Center: | Southern 242, what's your speed? |

TIME: 4:07:57

(Power interruption lasts for 36 seconds. Power restored at 4:08:33 PM at which point sound of rain is heard again on the cockpit radio.)

| | |
|---|---|
| Atlanta Center: | Southern 242, Atlanta. |
| Co-Pilot (Intra-C): | Got it, got it back Bill, got it back. |
| P/A System, Stdess B.: (Intra-cockpit) | Check to see that all carry-on baggage is stowed completely underneath the seat in front of you, all carry-on baggage, put all carry-on baggage underneath the seat in front of you. In the unlikely event that there is need for an emergency landing, we do ask that you please grab your ankles, I will scream from the rear of the aircraft, there is nothing to be alarmed but we have lost temporary power at times, so in the event there is an unlikely need for an emergency you do hear us holler, please grab your ankles, thank you for your cooperation and just relax, these are precautionary measures only. |
| Captain: | [to controller] Two forty two, stand by. |
| Atlanta Center: | Say again. |
| Captain: | Stand by. |
| Atlanta Center: | Roger, maintain one five thousand if you understand me, maintain one five thousand Southern 242. |
| Captain: | We're trying to get it up there. |
| Atlanta Center: | Roger. |

TIME: 4:09:15

| | |
|---|---|
| Captain: | Okay, [this is] 242, we just got our windshield busted and uh, we'll try to get it back up to fifteen [thousand feet], we're fourteen. |

| | |
|---|---|
| Atlanta Center: | Southern 242, you say you're at fourteen now? |
| Captain: | Yea—couldn't help it. |
| Atlanta Center: | That's okay. |
| Captain: | Our left engine just cut out. |
| Atlanta Center: | Say you lost an engine and busted a windshield? |
| Captain: | Yes, sir. |
| Atlanta Center: | Southern 242, you can descend and maintain one three thousand now, that'll get you down a little lower. |
| Co-Pilot (Intra-C): | My #### [deleted] the other engine's going to #### [deleted]. |
| Captain: | [to controller] Got the other engine going too. |
| Atlanta Center: | Southern 242, say again. |
| Captain: | Stand by—we lost *both* engines. |
| Co-Pilot (Intra-C): | All right Bill, get us a vector [heading] to a clear area. |
| Captain: | [to controller] Get us a vector to a clear area, Atlanta. |
| Atlanta Center: | Continue present southeastern bound heading, a TWA plane is off to your left about 14 miles at 14 thousand and says he's in the clear. |
| Captain: | Okay. Want us to turn left? |
| Atlanta Center: | Southern 242, contact approach control . . . and they'll try to get you straight into Dobbins [airfield]. |
| Co-Pilot (Intra-C): | I'm familiar with Dobbins, tell them to give me a vector to Dobbins if they're clear. |
| Captain: | [to controller] Give me vector to Dobbins if they're clear. |

TIME: 4:10:56
(Power interruption for 2 minutes and 4 seconds.)

| | |
|---|---|
| Captain (Intra-C): | There we go. |
| Co-Pilot (Intra-C): | Get us a vector to Dobbins. |
| Captain: | Atlanta, you read Southern 242? |
| Atlanta Approach Control: | Southern 242, [this is] Atlanta Approach control, go ahead. |
| Captain: | We've lost both engines—how about giving us a vector to the nearest place, we're at seven thousand feet. |
| P/A System Stwdess B (Intra-C): | Ladies and gentlemen, please check that your seatbelts are securely again across your pelvis area on your hips. |
| Co-Pilot (Intra-C): | What's Dobbins weather, Bill? How far is it? How far is it? |
| Atlanta App. Control: | Southern 242, roger, turn right heading one zero zero, your position is fifteen, correction twenty miles west of Dobbins at this time. |
| Captain: | Okay . . . twenty miles. |
| Atlanta App. Control: | Southern 242, right turn to one two zero. |
| Captain: | Okay, right turn to one two zero. |
| Co-Pilot: | [to controller] All right, listen, we've lost both engines and, uh, I can't tell you the implications of this, we only got two engines, and how far is Dobbins now? |
| Atlanta App. Control: | Southern 242, nineteen miles. |
| Captain: | Okay. |
| Atlanta App. Control: | Southern 242, do you have one engine running now? |
| Captain: | Negative, no engines. |
| Atlanta App. Control: | Roger. |
| Captain (Intra-C): | Just don't stall this thing out. |
| Co-Pilot (Intra-C): | No, I won't. |
| Co-Pilot (Intra-C): | What's the Dobbins weather? |
| Captain: | [to controller] What's your Dobbins weather? |
| Atlanta App. Control: | Stand by. |

| | |
|---|---|
| Co-Pilot (Intra-C): | Get [find] Dobbins on the approach plate [pilot's map]. |
| Captain (Intra-C): | I can't find Dobbins. Tell me where it's at? Atlanta? |
| Co-Pilot (Intra-C): | Yes. |
| Captain: | [to controller] Okay, we're down to forty-six hundred [feet] now. |
| Co-Pilot (Intra-C): | How far is it? How far is it? |
| Atlanta App. Control: | Roger, and you're approximately seventeen miles west of Dobbins at this time. |
| Captain: | [to controller] I don't know whether we can make that or not. |
| Atlanta App. Control: | Roger. |
| Co-Pilot (Intra-C): | Ask him if there is anything between here and Dobbins? |
| Captain (Intra-C): | What? |
| Co-Pilot (Intra-C): | Ask him if there is anything between here and Dobbins? |
| Captain: | [to controller] Is there any airport between our position and Dobbins? |

TIME: 4:16:28

| | |
|---|---|
| Aircraft intercom: | (Sound of three chimes.) |
| Stewardess A in forward Cabin (Int-C): | Sandy. |
| Stewardess B in rear cabin (Int-C): | Yes. |
| Stewardess A (Int-C): | They would not talk to me—when I looked in, the whole front windshield is cracked. So what do we do? |
| St. B (Intra-C): | Have they said anything? |
| St. A (Intra-C): | He screamed at me when I opened the door—just sit down—so I didn't ask him a thing, I don't know the results or anything, I'm sure we decompressed. |
| St B (Intra-C): | Ah yes, we've lost an engine. |
| St A (Intra-C): | I thought so. |

| | |
|---|---|
| St B (Intra-C): | Okay, Katty, have you briefed all your passengers in the front? |
| St A (Intra-C): | Yes, I told them I checked the cockpit— and help me take the door down. |
| St B (Intra-C): | Have you removed your shoes? |
| St A (Intra-C): | No I haven't. |
| St B (Intra-C): | Take off your shoes, be sure to stow them somewhere right down in the galley in a compartment in there with the napkins or something. |
| St A (Intra-C): | I got them behind the seat, so that's no good. |
| St B (Intra-C): | It might keep the seat down now. |
| St A (Intra-C): | Okay. |
| St B (Intra-C): | Right down in one of those closets, I took off my socks so I'd have more ground pull with my toes, okay? |
| St A (Intra-C): | You'd have what? |
| St B (Intra-C): | So I took off my socks so I wouldn't be sliding. |
| St A (Intra-C): | Yea. |
| St B (Intra-C): | Okay. |
| St A (Intra-C): | That's a good idea too. |
| St B (Intra-C): | Okay. |
| St A (Intra-C): | Thank you, bye bye. |
| Atlanta App. Control: | Southern 242, no sir, closest airport is Dobbins. |
| Captain: | I doubt we're going to make it, but we're trying everything to get something started. |
| Atlanta App. Control: | Roger, well there is Cartersville, you're approximately ten miles south of Cartersville, fifteen miles west of Dobbins. |
| Co-Pilot (Intra-C): | We'll take a vector to that, yes, we'll have to go there. |
| Captain: | [to controller] Can you give us a vector to Cartersville? |

| | |
|---|---|
| Atlanta App. Control: | All right, turn left, heading of three six zero, direct vector to Cartersville. |
| Captain: | Three six zero, roger. |
| Co-Pilot (Intra-C): | What runways? What's the heading on the runway? |
| Captain: | [to controller] What's the runway heading? |
| Atlanta App. Control: | Stand by. |
| Captain: | And how long is it? |
| Atlanta App. Control: | Stand by. |

### TIME: 4:17:08

| | |
|---|---|
| Captain (Intra-C): | Like we are, I'm picking out a clear field. |
| Co-Pilot (Intra-C): | Bill, you've got to find me a highway. |
| Captain (Intra-C): | Let's get the next clear open field. |
| Co-Pilot (Intra-C): | No #### [deleted]. |
| Captain (Intra-C): | See a highway over—no cars. |
| Co-Pilot (Intra-C): | Right there, is that straight? |
| Captain (Intra-C): | No. |
| Co-Pilot (Intra-C): | We'll have to take it. |
| Atlanta App. Control: | Southern 242, the runway configuration—at Cartersville is three six zero and running north and south . . . and uh, trying to get the length now—it's three thousand two hundred feet long. |
| Captain: | [to controller] We're putting it on the highway, we're down to nothing [in altitude]. |

### TIME: 4:18:07
(All of the following transmissions are intra-cockpit)

| | |
|---|---|
| Co-Pilot: | Flaps. |
| Captain: | There at fifty. |
| Co-Pilot: | Oh #### [deleted]. Bill, I hope we can do it. |
| Co-Pilot: | I've got it, I got it. |
| Co-Pilot: | I'm going to land over that guy. |

| | |
|---|---|
| Captain: | There's a car ahead. |
| Co-Pilot: | I got it Bill, I've got it now, I got it [he has taken the controls and is flying the aircraft]. |
| Captain: | Okay. |
| Captain: | Don't stall it. |
| Co-Pilot: | We're going to do it right here. |
| Woman's voice: | Bend down and grab your ankles. |
| Co-Pilot: | I got it. |
| Cockpit radio: | (Sound of breakup) |
| Unknown voice: | (Unintelligible) |
| Cockpit radio: | (More breakup sounds) |

End of tape
TIME 4:18:43 PM

# Chapter 12

# Fire in the Air

In the last chapter, the airliner was seeking to conserve altitude in its glide, in order to try to reach a runway on which it could safely touch down. In this chapter, the flight crew did the opposite. They deliberately headed for the ground as quickly as possible because they were faced with the most dreaded situation pilots can experience while flying—a fire in the aircraft.

The possibility of fire in the air has always been a source of concern for aviation professionals as well as passengers. However, strange as it may seem, from the beginning of passenger airline travel up until June 1983, there has never been an in-flight commercial aircraft fire in the United States that has ever resulted in a fatality. Unfortunately, this marvelous record could not last. On June 2, 1983, an Air Canada DC-9 jet, carrying 41 passengers and five crewmembers on the way to Toronto, Canada, finally, and reluctantly became the first aircraft to break that record.

"We have a fire in the back washroom," radioed the pilot to the control tower of Greater Cincinnati Airport in Covington, Kentucky. "And," he continued, ". . . we're filling up . . . with smoke right now. . . ."

The fire had apparently started in a rear lavatory from an electrical short in the wires of a toilet-flushing motor. Initially, as a steward battled the blaze with a fire extinguisher. the fire was thought to be under control. However, when the contents of two extinguishers were emptied without effect, the First Officer told the Captain, "I think we better go down."

From that moment on (7:06 PM), it was a race against time. As smoked started to billow towards the front of the aircraft, the pilot requested directions and emergency landing instructions to the nearest airport. He also transmitted, "Advise people on the ground we're gonna need fire trucks."

Replied the controller, "The trucks are standing by for you, Air Canada, can you give me the number of people [on board] and amount of fuel?" The response was, "We don't have time, it's getting worse here." The controller acknowledged, "Understand, sir. Turn left now, and you're just a half a mile north of final approach course."

As the smoke in the cabin thickened, passengers were forced to cover their faces with cloth or tissu₂ to act as filters (oxygen masks were not distributed by the crew because of concern that they might fuel the fire). At one point, as black smoke rolled into the cockpit, even the Captain's visibility became almost non-existent. "Where's the airport?" was the radio query to the controller.

Finally guided down to the runway, the aircraft made its landing only 13 minutes after starting its emergency descent. As the airplane came to a stop, flames were seen bursting out all over the fuselage.

Half of the occupants were able to escape within 30 seconds. The remaining 23 (all passengers), despite the dedicated efforts of the crew, were unable to reach the emergency exits before smoke or flames killed them.

Said a doctor who treated survivors at the hospital, "There was a very fine borderline between life and death. . . ." Another physician, helping remove bodies from the aircraft, agreed that only 30 seconds made a major difference. He said that many of those who perished had been, as he phrased it, "so close" to surviving.

The National Transportation Safety Board, after its investigation, issued a report which was contradictory in that it was both critical and complimentary of the actions of the flight crew. On the one hand, the Board felt that although it was reasonable for the crew members to initially take precious minutes to assess the extent of the fire in the washroom, a second delay based on a falsely optimistic report that the fire appeared to be abating, ". . . contributed to the severity of the accident." A decision to descend after the first inspection would have placed the aircraft on the ground three to five minutes earlier.

In addition to the delay, the decision of the First Officer to turn off the air conditioning because he believed the circulating air was ". . . feeding the fire," accelerated the accumulation of heat and toxicity in the cabin, decreasing the ability of the passengers to safely evacuate the aircraft.

On the other hand, the Safety Board termed the Captain's control of the aircraft during the descent an "outstanding" exhibition of airmanship, "without which the airplane and everyone on board would have certainly perished."

Chapter 13

# On a Collision Course
# With the Tallest Skyscraper

Most of the accidents that we've covered so far have been due to crashes with the terrain, or with other aircraft (in the air or on the ground).

In this chapter, we will cover the potential and actual encounters of aircraft with what appears to be a proliferating and perpetual hazard—skyscrapers located near airport landing and takeoff patterns, thrusting their heads through the clouds and up into the sky.

In July of 1945, on a foggy Saturday morning, a twin-engine B-25 military airplane (vintage World War II), crashed into the upper floors of the tallest structure in New York at the time, the Empire State Building. Despite the fact that the B-25 is a relatively small aircraft, the force of the impact was such that parts of the airplane hurtled clear across and through the 78th floor exiting the building from the other side, while the exploding fuel set fire to everything combustible from the 75th to the 79th floors. A total of 13 people were killed (including the three occupants of the airplane) and 26 injured either in the building itself or on the streets below from the shower of debris from above.

Now, use your imagination and visualize what could happen if a much larger and heavier modern jet airliner, with 58 persons on board, carrying considerably more fuel in the tanks, flying at twice the speed of a B-25, crashed into the upper floors of the much taller 110 story World Trade Center in New York. That almost happened in 1981.

On the 20th day of February that year, an Argentine airliner, Aerolineas Argentinas Flight 342, had made a stop at Miami Airport in Florida and was now on its way to John F. Kennedy Airport in New York. The 707 jet, heading for its final destination in the United States before returning to Argentina, carried a crew of nine and 49 passengers.

The Argentine aircraft experienced some holding delays due to weather conditions in the New York area. The ceilings were low, with heavy fog, and the sky was mostly obscured. Due to the minimum visibility, the approach delays to Kennedy Airport amounted to 90 minutes or so. Air controllers were quite busy giving instructions to airplanes they were tracking and directing via radar.

The New York controllers operated out of a modern building in Hempstead, New York. This brand new facility serves as a tracking center called TRACON, which stands for Terminal Radar Control. As many as 40,000 flights a day in the New York area can be handled through this up-to-date radar station.

Flight AR342, the Argentine airplane, was one of eight aircraft assigned to a controller handling traffic at Kennedy that evening. (Traffic was heavy and other controllers had their share of aircraft to track.)

Following are the recorded transmissions between the controller and AR342 (radio conversations between the busy controller and other assigned aircraft were omitted for the sake of clarity):

| | |
|---|---|
| Controller: | Argentine 342, turn right, heading zero four zero [degrees], descent to two thousand seven hundred [feet]. |
| AR342: | [Acknowledges] right to zero four zero, will you repeat the altitude? |
| Controller: | Argentine 342, RVR [runway visual range which is the length of the runway visible to the pilot] is two thousand eight hundred [feet] can you make the approach? [Note: altitude was not repeated] |
| AR342: | Will you say again the restriction for Argentine? |
| Controller: | Ah, yes sir, runway one three left [runway to land on] visual range two thousand eight hundred, is that sufficient for you to make the approach? |
| AR342: | Good, thank you [Note: Argentine 342 did not again ask for his assigned altitude] |
| Controller: | OK. |

From this moment on, the controller with the responsibility of tracking AR342, became busy giving instructions to other aircraft assigned to him. As far as he was concerned, he had assigned an altitude of 2,700 feet to AR342 that he expected to be maintained. However, the Argentine pilot, who asked for confirmation from the controller of the new altitude, did not receive an answer (possibly due to radio interference) and, unfortunately, did not pursue his request for clarification of the assigned altitude. (At a later hearing, when the controller was asked if there was any language problem in the communications with the Aerolineas Argentinas pilot, he answered, "Yes, on a scale of one to ten of understanding—least to best—the level was about four.")

This, then, is what occurred so far. The controller had given altitude and heading instructions and expected AR342 to maintain 2,700 feet. However, the pilot of Argentine Flight 342, flying at approximately 230 miles per hour, evidently had not heard or misinterpreted the 2,700 foot altitude assigned to him, and descended below that altitude to 1,500 feet.

Unbeknownst to the pilot flying through thick clouds, directly in the path of the Argentine airplane was the 110-story twin towers of the World Trade Center which, together with its steel TV mast, stood 250 feet *higher* than the altitude the Argentine aircraft had descended to!

We now continue the tape-recorded transmissions between Aerolineas Argentinas Flight 342 and the controller in the radar tracking station with the realization that there had been no radio contact between them for a relatively long period of time—two minutes and 47 seconds. During that period, the controller was directing other aircraft while the Argentine airplane descended from 3,000 feet, to and through the recently-assigned 2,700 feet, and all the way down to his present altitude of 1,500 feet.

(Just prior to resuming the transmissions, we can visualize the controller mildly eyeing the blip of the Argentine aircraft flying directly towards the World Trade Center complex at an altitude the controller believed was safely over that gigantic obstacle— until the Low Altitude Alarm went off.)

| | | |
|---|---|---|
| 10:06:32 PM | | [Sound of alarm buzzer at radar tracking center indicating Low Altitude Alert.] |
| 10:06:37 | Controller: | Argentine 342, what's your altitude? |
| 10:06:40 | AR342: | One thousand five hundred. |
| 10:06:41 | | [Sound of alarm buzzer at radar tracking center.] |
| 10:06:42 | Controller: | Altitude is WHAT?! |
| 10:06:44 | AR342: | One thousand five hundred. |
| 10:06:46 | | [Sound of alarm buzzer at radar tracking center.] |

A written report later submitted by the controller team supervisor in charge of the Kennedy sector that evening stated, "I was standing by the radar position, when I observed a 'Low Alt' alert on one of our inbound flights. It was Argentine three forty-two [AR342] and it appeared that he was at 1500 feet. AR342 was about 5 or 6 miles southwest of the World Trade Center [which is 1,749 feet high] and heading northeast right at that building. I immediately asked the controller, 'What's his altitude?' I walked over to the controller who was talking to AR342, and heard him ask Argentine, 'What's your altitude?' I didn't hear Argentine's answer, but the controller appeared startled and told Argentine to 'Turn right immediately!' "

| 10:06:46 | Controller | Argentine 342, turn right, immediate right turn, heading one eight zero [degrees]. |
| 10:06:49 | AR342: | Right one eight zero, Argentine. |

Continued the controller team supervisor in his submitted written report, "I yelled out, 'Climb him, climb him.' "

| 10:06:51 | | [Sound of ALARM buzzer at radar tracking center indicating Low Altitude Alert.] |
| 10:06:51 | Controller: | Argentine 342, climb, climb *immediately,* maintain three thousand [feet]. |
| 10:06:57 | AR342: | [Acknowledged] Climb to, climb to three thousand, Argentine. |

Since it does take time for a 707 jet at the slower approach speed to react to the commands of a pilot both in turning and climbing, and since the Argentine aircraft's present altitude and heading could be measured in seconds away from crashing into the 110-story World Trade Center north tower, it was with bated breath that the controller and his supervisor watched the blip on the radar screen slowly begin to turn and climb away from the tallest building in New York City.

The Supervisor's written report then continued with a matter-of-fact and rather anticlimactic statement, "I observed the Argentine 707 turn and climb and he missed the World Trade Center. . . ."

Continuing the tape-recorded radio transmissions:

| 10:06:57 | AR342: | I'm pretty sure now [my] read back to you [was] descent to one thousand five hundred [feet]. |
| 10:07:09 | AR342 | I'm visual over the clouds passing two thousand up to three thousand, Argentine. |
| 10:07:14 | Controller: | Argentine 342, [continue to] climb and maintain three thousand [feet], fly heading of one eight zero [degrees]. |
| 10:07:19 | AR342 | [Acknowledges] One eight zero passing [through] two five thousand [error—he means two five hundred]. |
| 10:07:36 | AR342 | Three forty-two reaching three thousand. |
| 10:07:39 | Controller: | Argentine 342, roger. I gave you a speed of a hundred eighty [knots] and [an altitude] of twenty seven, two thousand seven hundred feet. Maintain three thousand for now. I'll take you out [vector you] for another approach. |
| 10:07:48 | AR342: | Okay, sorry sir. I understood one thousand five hundred and read back [to] you. Sorry so much. |
| 10:07:52 | Controller: | Okay, I missed the read back, apparently. Ah . . . there was a building up there ahead of you that was quite high. |
| 10:07:59 | AR342: | Okay, don't worry for me, thank you. |

The controller team supervisor concluded his written report, "AR342 was resequenced and landed without further incident. I reported this incident . . . to the Comm. Center. Then I wrote a Flight Assist [commendation] for the controller's quick action. Then I wrote a Pilot Deviation on the Argentine aircraft."

An FAA official noted the pilot didn't have perfect command of

English and pointed out the fact that the air traffic controller had a heavy workload on that poor visibility night. "The pilot is Argentine and he doesn't speak good English, as you can tell from the tape," said the TRACON Air Traffic chief. "He obviously cranked in the wrong altitude and the controller was very busy. He had seven or eight planes on the same frequency."

The controller, who handled the Argentine aircraft along with seven other airplanes assigned to him, submitted a brief written report on what was now called the World Trade Center Incident. In addition, he was asked to appear at a hearing held at the New York tracking center where he was asked the following questions:

Question:     Could he expand his written statement?
Answer:        He felt the pilot did an excellent job of turning and climbing as requested.
Q.               Were any altitude clarifications made by AR342 ever heard?
A.               No, he never heard any acknowledgement of anything less than 3,000 feet.
Q.               When did he become aware of the low altitude?
A.               Just before the MSAW (Minimum Safe Altitude Warning) alert sounded.
Q.               What role did the MSAW alert play?
A.               He never heard the alert before he called AR342. It apparently sounded just as he called AR342.
Q.               Any idea of how AR342 could remain off an assigned altitude for so long without being noticed?
A.               He didn't know.

Evidently, the controller recognized the danger to the Argentine aircraft at about the same time the MSAW alarm sounded. He was so shaken up after the incident that he went on sick leave. When interviewed at his home a week later, he explained, "When something goes wrong it scares the life out of you. It was the shock of my life. It still hasn't worn off," he said. "There's been close ones before but this one weighs on me. It lingers."

He added, "I don't know how it got that . . . low. But, I felt that the pilot did a heck of a job. A 707 with a ground speed of

The view from a graveyard in Scotland – a section of the wreckage of a Pan Am Boeing
747 brought down by a terrorist bomb in December 1988.

The devastation of Sherwood Crescent, Lockerbie, which took the full force of the downed 747.

A body on a roof, a seat embedded in a window – two shocking sights in Sherwood Crescent.

The cockpit section of flight 103.

The wreckage of the British Midland Boeing 737 which crashed onto the embankment of the M1 in January 1989.

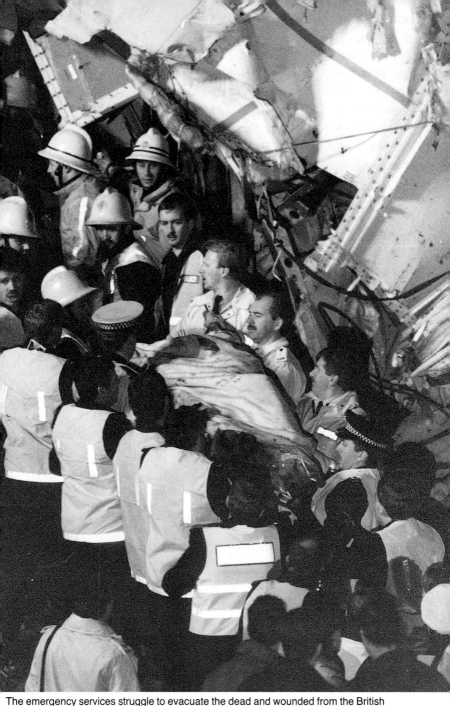

The emergency services struggle to evacuate the dead and wounded from the British Midland disaster.

An aerial view of the M1 crash which shows just how close Captain Hunt came to nursing his injured plane onto the runway of East Midlands airport.

British Midland Flight BD92 lies broken-backed between the two carriage ways of the M1.

A view of the tail section of Flight BD92, overhead another aircraft continues its descent into East Midlands airport.

240 knots [over 275 miles per hour], does not react quickly."

He stated that his doctor would determine when he could return to work. "I've been a controller for 14 years and its the only thing that I want to do," he said. "In my case, a traumatic injury could be considered more of a mental injury than a physical injury. I've been under doctor's care since . . . the day after the incident," he explained. "Every controller has things that happen to him like this. But this is the first time I've reacted like this. My doctor told me it's a buildup, an accumulation."

The president of the Professional Air Traffic Controllers Organization's local union (PATCO) described the reaction as normal for a controller involved in this type of situation. "You have to understand," he said, "that in his mind, as he took immediate action to turn the airplane, there was the question of whether the plane would turn in time and whether the pilot would react to his instructions quickly enough to avoid an accident."

According to the local PATCO president, the controller's reaction was extremely fast considering the fact that he was working three radar screens and controlling seven or eight aircraft, the night he saved New York from a catastrophe. In addition, he was controlling the most dangerous approach to Kennedy Airport—but the only one that could be used in poor visibility that night. "I think that the mayor should give that guy the keys to New York City," he said.

An official of the Federal Aviation Administration stated that the FAA may give the controller a citation for his actions in averting a disaster. A local New York newspaper wrote in an editorial, "We owe [the controller] a medal." However, when the FAA chief of air traffic operations for the northeast was asked whether the controller could be designated a hero, he replied, "No, not in a true sense of the word—he did what he was supposed to do."

The air traffic operations chief said that there was no question that the Argentine aircraft would have hit the World Trade Center if it had not been warned away. "No doubt about it," he said. "Especially since he was on his way down to 1,400 feet. It looked like he was headed right toward the north tower, or the west corner of it. If the controller hadn't caught it, if the system hadn't sounded," he added, "well, that's where he was headed."

The MSAW (Minimum Safe Altitude Warning) system was fairly new, having been installed in the New York radar tracking facility for less than a year. It operates with the help of a computer which contains data on the height of the highest terrain or man-made structures in the area. When radar spots an aircraft within 500 feet of the highest obstruction and 30 seconds away, a warning buzzer sounds repeatedly in the radar tracking center.

The incident involving Aerolineas Argentinas turned out to be the first occasion that the MSAW system was used effectively. "It just paid for itself with that one time," said the chief of TRACON (Terminal Radar Control).

Another FAA official heartily agreed. "The people who should be glad for this system are one Argentine pilot and probably a lot of people in lower Manhattan. He would have hit the building if [the new alarm system] wasn't in operation."

Despite the high marks given the new MSAW device by the FAA, the president of the local controller's union claimed that the system is so unreliable that the air controllers have no faith in it. "The FAA wouldn't tell you, but it frequently goes off when aircraft are [as high as] 15,000 feet. It's very distracting and controllers often have to override it," he said.

An FAA investigating team met with Aerolineas Argentinas representatives a few weeks after the "near-miss" and the latter offered to present both of the Argentine pilots for interviews.

Along with the oral statements by the pilots, the written investigative report contained this interesting paragraph: "The Captain asked . . . if the FAA had called in reference to his flight. He stated he was fully prepared to preserve the cockpit voice recorder [CVR] data to aid in any FAA inquiry. Since no FAA request was made, the CVR data is not available [was erased]." (Too bad. The tape of the cockpit voice recorder would have probably answered a number of key questions.)

There was nothing that the FAA could do in this situation. However, the hazardous incident involving Aerolineas Argentinas may have drawn special attention to it from the FAA. This was the second "almost disaster" in the last two years and the third one involving that airline in the last four years.

The earlier "near catastrophe" concerned an Argentine 707 jet that took off from Kennedy Airport in a driving snowstorm skimming rooftops in the Woodmere-Cedarhurst, Long Island, area. A number of terrified residents literally dove to the floor as the jet aircraft came within 100 feet of smashing into the heavily populated area.

The other almost fatal disaster by an Argentine pilot took place a year earlier. He made too sharp a left turn after taking off from Kennedy Airport and began overtaking a Pan Am jumbo jet ahead of him. When the controller radioed traffic instructions to the Pan Am to make a turn to get out of the way of the following aircraft, the Argentine pilot thought this change of direction was for him and followed suit, still overtaking the Pan Am. The controller urgently issued new directional orders to the Pan Am, but again, the Argentine pilot misunderstood and followed instructions meant for the other aircraft. Finally, in desperation, the controller radioed, "Argentine 431, I want you to listen closely!" This time, the Argentine pilot understood. Another potential collision averted—just in time.

Based on the fact that these three hazardous incidents in the New York area involved Argentine aircraft, a Congressman from New York sent a letter to the Federal Aviation Administration asking them to examine the credentials of pilots flying for that airline. He said, "Three 'near misses' by one airline seems extraordinary. I don't believe we should sit back and wait for a catastrophe before taking action. If there is some problem with Aerolineas Argentinas, it must be corrected if that airline is to continue to fly into U.S. airspace."

In summing up the Argentine incident involving the "near-collision" with the World Trade Center, the negatives could be listed as follows:

1. The pilot did have some difficulty with the English language (which is the common language for all international pilots).

2. Although the pilot did ask for clarification of his assigned altitude, he never repeated his request when it wasn't furnished.

3. The controller was too busy to ask for confirmation of the altitude assigned and assumed the pilot received and understood

the instructions at the time they were issued.

4.   In the interest of safety, was it wise to issue a compass heading to the 707 jet airliner where it would have to pass over the tallest buildings in New York, the 110-story twin towers of the World Trade Center? There's always the possibility that an aircraft's altimeter may be inaccurate or have been set improperly.

On the positive side:

1.   The new MSAW system did perform as designed and alerted the TRACON unit to the danger of an aircraft at too low an altitude.

2.   The controller gave instructions quickly and decisively, once made aware of the altitude of the Argentine aircraft.

3.   The controller supervisor was also alert and responded promptly and efficiently in backing up the controller.

4.   The Argentine Captain immediately complied with the new radio commands issued by the controller and skillfully piloted his aircraft up and away to a safe altitude. Result? What could have been recorded, in the history of aviation crashes, as one of the more devastating accidents, was finally officially classified by the FAA as only an incident.

---

Chapter 14

# Engine Falls Off on Take-Off

It was nice to have concluded a chapter in which an accident, although very close to becoming an extraordinary newsworthy event, actually did not occur. Unfortunately, a near-miss of this magnitude is more the exception than the rule.

Now you will read about the fatal take-off of an airliner that received much media attention. Despite the extensive publicity, the inside story of the accident was not easy to discover.

This one was due to equipment failure. Not something minor that you've already read about, such as a blow-out of a tire or a malfunctioning gear light; this one involved a major structural failure occurring at a critical stage of takeoff. Regrettably, the termination of this flight was not a near-miss.

On May 25, 1979, 258 passengers boarded a DC-10 wide bodied aircraft at Chicago-O'Hare International Airport, Illinois. This was American Airlines Flight 191, a regularly scheduled passenger flight bound for Los Angeles, California. Together with the 13 crew members (Captain, First Officer, engineer and 10 flight attendants), there was a total of 271 persons on board for this flight. A lot of people on one airplane.

Flight 191 started to taxi from the gate just before three PM Central Standard Time. Maintenance persons who monitored the flight's engine start, push-back (away from the gate) by tow-vehicle, and the start of the taxi, did not observe anything out of the ordinary. No one could possibly suspect at the time that in less than five minutes, every single one of the passengers and crew members aboard that tranquil-looking DC-10 would die a horrible death.

The weather at the time of departure was clear, with excellent visibility. The aircraft was given permission to begin its taxi to the runway in preparation for the takeoff. The gross weight of the airplane was computed to be 379,000 pounds, a rather heavy load.

The jumbo airliner was then instructed to taxi into position on the runway and hold, waiting its turn to take off. A minute later, the aircraft was cleared for the takeoff and, shortly after, the Captain acknowledged to the tower, "American one ninety-one under way." The Captain, 53 years old, was a seasoned veteran with American Airlines with a considerable number of hours of flying experience, many of which were as Captain in the DC-10.

The takeoff roll of the three-engine jet airliner appeared to be normal. However, just before Flight 191 reached the rotation point on the runway where the controls are pulled back to lift the nose of the aircraft (the speed was then 167 miles per hour), sections of the number one (left) engine pylon structure (the housing encasing the engine) started to come off the aircraft. White smoke started to stream from the vicinity of the number one pylon. A second later, the word "damn" was recorded on the cockpit voice recorder.

During the rotation a bizarre event took place. The number one engine together with its pylon assembly actually tore itself loose from the aircraft. The engine and assembly traveled up and over the left wing without striking any of the critical aerodynamic or flight control surfaces of the airplane before smashing to the ground behind the accelerating airliner. Thousands of pounds of engine and pylon assembly parts were left scattered behind on the runway.

The air traffic controller in the tower, observing the takeoff roll of Flight 191, saw the left engine break off at the point of liftoff. He immediately transmitted, "American 191, do you want to come back? If so, what runway do you want?" There was no reply. Apparently, the flight crew members were having their hands full at the time.

Despite the sudden and unexpected loss of the number one engine at this stage of rotation, the flight crew was committed to continuing the takeoff. (Authorities later stated that the flight crew's decision not to attempt to discontinue takeoff "was in accordance with prescribed procedures and was logical and proper in light of information available to them.") The aircraft reached the takeoff speed of 183 miles per hour and became airborne.

Flight 191, with the left engine gone, lifted off in a slight left wing-down attitude. The First Officer, also an experienced pilot with many hours of flying the DC-10, was at the controls of the aircraft. He restored the airplane to a wings-level position and maintained a steady climb at a 14 degree noseup pitch attitude—which is exactly what is called for on a climb-out for a DC-10 aircraft with only two of its three engines operating.

Nine seconds after liftoff, the jumbo aircraft reached the height of 140 feet above ground and was climbing at the speed of 172 knots (198 miles per hour). The pitchup attitude of 14 degrees noseup, as well as the headings, remained fairly stable and it appeared that the aircraft, despite the loss of one of its three powerful engines, was responding well to the flight crew's controls.

A photographer, who holds a pilot license, was at the airport that afternoon. He not only witnessed the takeoff but actually took pictures of the DC-10 in flight.

"I was in the terminal lounge looking out towards the runways and I watched the DC-10 right from takeoff. It rolled away and accelerated down the runway and it lifted off and started to climb. At this point, nothing looked wrong; everything was going just as it should."

He was holding his camera with a telephoto lens ready to shoot a picture of another aircraft, when his attention was brought back to the DC-10. "The left engine seemed to explode away from the wing although there was no smoke or flame.

I saw the engine come tumbling through the air, tumbling and tumbling, to the ground. There was still no flame," he recounted, "but there was a stream of smoke or vaporizing fuel coming from the trailing edge of the wing where the engine used to be."

Flight 191 continued to climb with the nose up 14 degrees but the airspeed started to slowly decelerate. In ten seconds, the airspeed diminished from 172 knots to 159 knots and, in that interim, the airplane had climbed to the highest point it would ever reach on that ill-fated flight—350 feet above the ground.

At 172 knots, the climb-out was stable; at 159 knots, which turned out to be a critical airspeed for this wounded DC-10, the aircraft started to act up. (Had the nose of the airplane been lowered slightly, the 172-knot airspeed necessary to keep this damaged aircraft under control could have been maintained. However, the pilot was literally following the book procedure which called for a 12- to 20-degree noseup attitude as the best configuration on climb-out when the DC-10 had only two engines operating.) At 159 knots, the DC-10 began to lower the left wing and roll to the left.

(Later, a spokesman for the Federal Aviation Administration, in responding to reporters' questions, stated, "He could have flown that thing [the DC-10] all the way to Los Angeles [on two engines]. The problem was not engines—it was low altitude. He was at such a low altitude that he did not have enough room to take corrective action.")

The pilot applied right rudder (to steer the nose to the right) and right wing-down aileron (to lower the right wing). However, Flight 191 continued to roll and turn to the left despite increasing right rudder and right wing-down deflections. Within seconds, the aircraft was completely on its left side.

"Just after the engine was blown off was when I made the first picture," continued the photographer. "I took the second after the aircraft started banking very sharply to the left. It was, by that time, nearly perpendicular to the ground." The pilot-photographer added, "The aircraft had rolled to the left more than 90 degrees and was starting to roll onto its back. The nose was pointed down. . . ."

Despite the frantic and desperate application of full right aileron and right rudder controls, the aircraft continued to turn and reached a 112-degree left roll in a nosedown attitude. Now, nothing could stop the DC-10's headlong rush towards the earth.

At an airspeed of 183 miles per hour, Flight 191 flew directly into the ground in a left wing-down and nosedown attitude. The left wingtip hit first and the aircraft exploded, broke apart and was scattered, along with its occupants, into an open field and a trailer park. The disintegration was so extensive that, afterwards, it was almost impossible to identify the debris and endless bits of wreckage as that of an airplane.

The horrified photographer said that the DC-10 had dropped from his view behind the terminal building and the last thing he saw was a red ball of flame and black smoke. He snapped another picture as the bright orange flame rose into the sky. He concluded, "When I saw all that violence and flame, I knew nobody could have gotten away alive."

All of the 271 persons on board Flight 191 perished instantly. The official report stated that this accident "was not survivable because impact forces exceeded human tolerances."

In addition to the non-survivable impact forces, a huge ball of fire erupted from the explosion of the 60 tons of unused fuel which had been stored in the tanks of the DC-10. The raging fire succeeded in horribly burning all of the 271 crash victims.

The DC-10 crashed near the property of a man who was standing outside of his office building, oblivious to the airplanes landing and taking off at nearby O'Hare airport. The first he knew of the accident was when he heard an explosion. "By the time I looked up," he said, "there was a rain of fire falling down on me."

A Chicago policeman actually saw the airplane crash into the ground. "As soon as it went down, it went up in flames," he said. "Swish, just like napalm."

The pastor of Queens of the Rosary Catholic Church who arrived almost immediately after the tragic crash recalled, "It was too hot to really do anything but administer the last rites. I said some prayers and gave a general absolution. I just walked around trying to touch a body here or there, but I could not.

"It was too hot to touch anybody and I really could not tell if they were men or women. Bodies were scattered all over the field."

The head of the Federal Aviation Administration visited the crash scene. "It's overwhelming," he said. "It's hard to tell there was a DC-10 there. I'm sure the pieces add up to one, but . . ." His voice trailed off.

When he was asked to compare this disaster to other major air crashes, he noted, "The dimensions here were twice as large," he said. "This plane was so much bigger."

The force of the impact was hard to believe. "We didn't see one body intact," said a fireman trying to recover bodies from the wreckage vicinity. "Just bits and pieces. We haven't been able to see a face or anything—just trunks, hands, arms, heads and parts of legs.

"But we can't tell whether they were male or female, whether they were an adult or child, because they were all charred." Bits of human flesh and bones were on the ground and, in some cases, stuck to the wreckage.

Hands were being matched to arms, heads to necks, feet to legs and legs to torsos. "We just guessed," said a volunteer worker. "Everywhere we walk, there are bodies and sometimes we have stepped on a few."

People worked through the night by floodlights to remove all of the victims. It would take them into the following morning to complete the task. Meanwhile, parts of the DC-10 were still smoking and the smell of acrid fumes continued to permeate the air.

Bodies, with matching parts, were placed in rubber and plastic mortician's bags which were then carried to a police van by four men. At times, the vans found it difficult to keep up with the number of body bags awaiting removal. Finally, all of the victims were placed in an American Airlines hangar serving as a temporary morgue.

The Cook County Medical Examiner said that, in many cases, dental records would be required because human remains found were too badly burned or dismembered. "The situation is such that identification will be difficult," he said. In fact, "some bodies may never be identified." Especially since fingerprinting in a number of cases would be impossible.

He referred to the hangar morgue as a "cemetery without coffins." The black and yellow rubber and plastic mortician's bags were laid out on mattresses. Only about a dozen bodies out of the 273 victims were intact with upper torsos and heads.

The distraught Mayor of Chicago was on the scene offering her sympathy and sorrow for the victims and families. "I am sure that the citizens of the city join me in offering their prayers for these people. . . . We in Chicago will do all we can to continue to assist the families of crash victims . . . in any way possible."

The news of the crash seemed to be the only topic of conversation of the residents of Chicago that day. Some travelers planning to fly to Los Angeles within the next few days expressed great relief that they were not scheduled to fly on Flight 191 that ill-fated day. "I get nervous any time I fly," said one lady due to make that trip. "It could have been me."

Since the aircraft had crashed in Chicago within five minutes of takeoff, most of the relatives and friends planning to meet the flight at the Los Angeles Airport terminal learned about the horrible news from radio and TV reports hours before the scheduled arrival. These people, for the most part, realized it would be pointless to go to the airport and remained home awaiting further news developments.

However, approximately two dozen other relatives and friends were unaware of the tragic crash and arrived at the destination terminal as planned. These people, looking at the display screen in the terminal, which carries the arrival time of all inbound flights, saw a message next to Flight 191, "See Agent." Upon making inquiry, they were called aside by American Airlines personnel and gently given the bad news.

"It's so hard to tell people," said a red-eyed American Airlines employee. "We hate to tell them anything unless it's absolutely certain. We aren't in a position yet to tell anyone for sure whether their person was on the plane."

Each of the people close to the victims of the accident were then directed to a private room by American Airlines personnel. Unfortunately, although laden with grief, many of them had to force their way through a horde of reporters and photographers vying to get quotable statements or pictures.

"I parked my car out there where you're not supposed to," said one young man awaiting the arrival of his brother. "I heard about the crash on the radio just when I got here. I didn't know what to do," he said as he broke into tears.

In the private room, away from the terminal's bustling activity, a priest who happened to be at the airport, tried to minister to the grieving. Also, a physician was summoned by the airline to help those who required medical assistance. What added to the distress was the fact that American Airlines was temporarily having difficulty in compiling a complete and accurate passenger list of the disastrous flight.

"You are never really ready for something like this," said an American Airlines district manager. "We just try to help people as best as we can." When asked by a reporter how he reacted when he heard the news, he replied, "I got sick. I thought I was going to throw up."

Following normal procedure in accidents (especially of this magnitude), the National Transportation Safety Board immediately dispatched an investigative team to the scene. The Safety Board concentrated on two major aspects of this accident:

1.   Despite the loss of the left engine, why was the flight crew unable to safely fly the aircraft on the remaining two engines?

2.   What caused the pylon assembly to separate from the rest of the aircraft?

(If the reader will now bear with us, we will try to be brief and simplify what would ordinarily be a lengthy and technical explanation of the ultimate findings of the investigators.)

An examination of the wreckage revealed that a three-foot section of the wing attached to the pylon was also torn away when the pylon separated from the aircraft. This had an effect on the hydraulic system which not only made the stall warning system inoperative but caused an imbalance in the stability of the aircraft by retracting part of the left wing's slats. (Slats are panels that can be extended from each wing to provide additional lift at lower airspeeds. If part of one wing's slats are retracted while all the other slats remain extended, a stall and subsequent roll is apt to occur at too low an airspeed.)

A Douglas Simulator was used to duplicate the flight conditions of Flight 191 with the loss of the left engine, the retraction of part of the left wing's slats and the failure of the stall warning system to operate. (The stall warning device is known as a stick-shaker which alerts the crew to an impending stall by introducing a vibration into the pilot's control column.)

Thirteen pilots participated in the simulator tests. In all cases where these pilots duplicated the control inputs and pitch attitudes shown on Flight 191's flight data recorder (which was recovered from the wreckage), control of the aircraft was lost and the simulator followed the same path as Flight 191—into the ground.

Since each of the 13 pilots was thoroughly briefed on the fact that the jumbo jet started to behave erratically when the airspeed slowed down to 159 knots, they were all alert to any signs of the beginning of a left roll. When that started to occur, they immediately lowered the nose of the aircraft from the 14-degree pitchup attitude (thus increasing the speed), recovered from the left roll and continued the flight successfully. In those cases where the pilot attempted to raise the nose to regain the 14-degree pitchup attitude, the aircraft reentered the left roll.

When the simulator was programmed with an operating stick-shaker, it activated seven seconds after liftoff. At that point, the pilot increased his speed to 167 knots, safely over the critical 159 knots where the aircraft would start to roll. A stable climb was readily achieved at the steady airspeed of 167 knots (where the stick-shaker stopped vibrating) and the aircraft remained completely under control.

Pilots and test pilots who testified at the Safety Board's public hearing stated that flight crews cannot see the left engine and left wing from the cockpit and, therefore, the first warning the Flight 191 crew would have received of the stall was the beginning of the roll. None of the pilots believed that, under these circumstances, it was reasonable to expect the flight crew of Flight 191 to react in the same manner as did the simulator pilots. The latter had been made aware of Flight 191's loss of critical systems and thus were able to recover from the stall.

After reviewing the evidence, the National Transportation

Safety Board was able to come to the conclusion that ". . . the loss of control of the aircraft was caused by the combination of three events: the retraction of the left wing's outboard leading edge slats; the loss of the slat disagreement warning system; and the loss of the stall warning system—all resulting from the separation of the engine pylon assembly. Each by itself would not have caused a qualified flight crew to lose control of its aircraft, but together during a critical portion of flight, they created a situation which afforded the flight crew an inadequate opportunity to recognize and prevent the ensuing stall of the aircraft."

Having satisfied itself as to the reasons for the flight crew's inability to control the DC-10, the Safety Board now sought to determine why the left engine and pylon separated from the aircraft.

One of the tasks of the investigators was to painstakingly locate and collect all of the bits and pieces of the number one pylon assembly and the left engine found on, or alongside, the runway. The metal parts were then carefully examined at the Safety Board's metallurgical laboratory. There, engineers discovered a ten-inch long fracture of the part of the pylon that was attached to the aircraft. This fracture had all the characteristics of an overload separation and, unquestionably, resulted from over-stress.

As a result of this discovery, tests were conducted in an attempt to reproduce the ten-inch overload crack. These tests were carried out by both American Airlines and the DC-10 manufacturer, McDonnell-Douglas. Following the same attachments and connections used on the pylon of the accident DC-10, cracks appeared in the same place when the test metals were subjected to maximum engine thrust loads (which would occur at takeoff).

It was now logical for the investigators to examine the pylon assembly maintenance procedure followed by airline mechanics working on the DC-10.

The standard procedure called for removal of the engine and pylon separately when maintenance was necessary for that unit. However, two years before this fatal accident, American Airlines engineering personnel evaluated the feasibility of raising and lowering the engine and pylon assembly as a single unit using a forklift-type supporting device. This technique not only would save about 200 man-hours per aircraft, but also, from a safety viewpoint, would reduce the number of disconnects (hydraulic and fuel lines, electrical cables and wiring, etc.) from 79 operations to 27, thus reducing the error possibilities on re-connecting.

In checking with McDonnell-Douglas to the feasibility of using this cost-saving (as well as safety) feature, the response to American Airlines was that the aircraft manufacturer would not encourage this procedure due to the risk involved in the re-mating of the combined engine and pylon assembly to the wing attach points. However, since McDonnell Douglas does not have the authority to either approve or disapprove the maintenance procedures of its customers, American Airlines decided to go ahead and lower the engine pylon assembly as a single unit by means of a forklift.

What was learned after the disaster of Flight 191, was that because of the close fit between the pylon to wing attachment points, maintenance personnel had to be extraordinarily cautious while they detached and attached the pylon. A minor mistake by the forklift operator while adjusting the load, or even an imperceptible move of the fork arms, could easily damage the pylon assembly and ultimately result in a fracture due to overstress. Testimony of mechanics who performed this type of forklift maintenance confirmed that the procedure was rather difficult.

Three days after Flight 191's crash, the Federal Aviation Administration ordered an inspection of the pylon assemblies of the DC-10's in service. A total of nine DC-10's were discovered with fractures: four from American Airlines, one from United and four from Continental. The aviation authorities became concerned that many more DC-10's were damaged during maintenance and returned to flight duty.

Based on these preliminary findings, on June 6, 1979, the FAA issued an emergency order of suspension of flights of all DC-10's. Additional inspections were ordered and investigative teams were set up to determine the proper maintenance procedures to be followed in the future on the DC-10 aircraft.

After completing their investigation, the FAA issued orders mandating specific procedures to be followed to prevent possible damage to pylon assemblies during maintenance. In addition, orders were also issued outlining required inspection procedures for the pylon assemblies as well as for all slat system cables and pulleys.

It was not until July 13, 1979, that the suspension was lifted and the DC-10 was allowed to return to service.

In the aftermath of the tragedy of Flight 191, which turned out to be the worst domestic airline disaster in U.S. history, two separate ironic facts emerged:

1.   A victim of the Chicago DC-10 crash had experienced a personal tragedy 17 years earlier. Both of his parents had then been killed in an airplane accident in which, like Flight 191, all aboard perished. The airline was the same—American Airlines. The destination of their flight—the same, Los Angeles. A tragic coincidence.

2.   Like many of the other DC-10's, Flight 191 had a TV camera mounted behind the captain's shoulder to give passengers a cockpit view of the takeoff roll, liftoff and climb into the air. Passengers also had the ability, by using headsets distributed by the flight attendants, of listening to transmissions between the control tower and the DC-10 as well as other aircraft communicating with the tower. This system was set up to entertain the passengers by making them feel as if they were participating with the flight crew in controlling the aircraft. It is likely that, due to this closed-circuit TV, the passengers on Flight 191 actually saw themselves take off, climb—and then, nosedive into the ground.

(Note: Despite the fact that the DC-10 was cleared of any design flaws, orders for the aircraft collapsed after this latest tragedy. Shortly afterwards, Douglas Aircraft announced that it would no longer manufacture the DC-10 jumbo airliner.)

## Chapter 15

# Airliner Turned Completely Over on Its Back

In the last chapter, the pilot had been placed in the precarious position of battling the insistence of the aircraft to roll to the left, despite all his efforts to maintain level flight.

The next episode is similar to the Chicago accident in that the pilot was also unsuccessful in fighting the tendency of his airplane to roll. However, this time it was in the other direction—farther and farther to the right—to the extent where his passenger aircraft, at one stage of flight, wound up completely upside down and uncontrollable.

On April 4, 1979, Trans World Airlines Flight 841 departed John F. Kennedy Airport on a flight to the Minneapolis-St. Paul Airport, Minnesota. The Boeing 727 aircraft took off during the evening—at 8:25 PM—with 89 persons on board, 82 passengers and 7 crew members.

The takeoff, climb and en route portions of the flight were uneventful, until the airplane reached the altitude it had been cleared to, 35,000 feet. After flying at that level for a short while, the flight called the controller, stated that it was being slowed down by a headwind of over 115 miles per hour and requested permission to climb to 39,000 feet. Receiving the clearance to do so, the captain climbed the aircraft, leveled it at 39,000 feet and then engaged the autopilot in the altitude-hold position. (In this position, without any of the crew members touching the controls, the autopilot will make slight imperceptible adjustments of the controls of the aircraft to maintain the altitude and heading selected by the pilot.)

The Captain reported later that the visibility was excellent and that there was no turbulence. The airplane was flying smoothly, all systems were indicating normal operation and there were no warning lights showing on the panel.

However, while he was sorting maps and charts located in his flight bag on the floor of the cockpit, he felt a buzzing sensation. Within two or three seconds, the buzzing became a light buffet. He immediately raised his head to look around and noticed that the instruments and controls were not acting in concert. The auto-pilot was moving the control wheel to the left as if to make a left turn, but his instruments were indicating that the aircraft was actually banking to the right. (Since this was night-time with no visible horizon to use as a reference, a pilot has to depend upon his instruments to let him know whether his airplane is flying straight and level or turning.) When the aircraft continued to bank to the right at a faster rate of roll, he quickly disconnected the autopilot (the airplane could now be flown manually) and applied more left aileron control (which has the effect of lowering the left wing) to stop the roll to the right.

According to the Captain, the aircraft continued to roll to the right in spite of nearly full left aileron control, so he applied left rudder control (which ordinarily will steer the nose of the air-plane to the left) in addition to the aileron control. He said that, in spite of the almost full deflection of the left aileron and full displacement of the left rudder pedal, the aircraft continued to roll to the right.

All this was taking place as Flight 841 was flying at approxi-mately 300 miles an hour at an altitude of seven miles above Saginaw, Michigan. The Captain found himself fighting the determined tendency of the aircraft to roll more and more to the right despite all of his counteractive efforts. He desperately continued to apply full left aileron control and to fully depress the left rudder pedal. Nothing changed. The aircraft continued to roll farther and farther to the right.

Finally, realizing that the airplane's roll could not be stopped, he reduced his speed by retarding the throttles to the flight idle position, and cried out, "We're going over." At which point, Flight 841, with its 82 passengers and seven crewmembers, turned completely over on its back.

The aircraft, finding itself in this embarrassing (and dangerous) position, fortunately continued its roll to the right until it was once again right side up. But not for long. Now, it started a second roll.

This three-engine Boeing 727 jet airliner then proceeded to execute a series of acrobatic maneuvers certainly not designed or intended for a commercial passenger aircraft with 87 persons aboard. After having already completed a 360 degree barrel-roll, the start of another roll to the right was followed by the airplane lowering its nose to begin a breathtaking dive.

Flight 841 was now completely uncontrollable. Starting at the altitude of 39,000 feet, the flight angle quickly went from horizontal to a completely vertical downward position, with the aircraft picking up speed at it dived towards the earth. By the time the airplane plummeted to the 29,000-foot level, the airspeed needle had almost reached its limit. The Captain now extended the speed brakes, without effect, while continuing to maintain full left aileron and full left rudder, still trying to stop the rolling to the right. At the same time, he was applying back pressure to the controls to try to get the nose up, also without success.

Needless to say, the passengers were terrified. 39,000 feet . . . 35,000 feet . . . 30,000 feet . . . 25,000 feet . . . 20,000 feet . . . like a runaway roller coaster heading straight down, faster and faster and faster!

At 15,000 feet, the Captain ordered his co-pilot to lower the landing gear in an attempt to slow the descent of the aircraft. As the First Officer moved the gear handle to the "extend" position, the flight crew heard a loud sound similar to the sound of an explosion (the landing gear is not supposed to be lowered when the aircraft is flying at too high a speed for fear of damaging the gear and/or the aircraft). Whether coincidental or not, lowering the gear seemed to have a beneficial effect on the aircraft.

At last, the airspeed began to slow up a bit and the captain could feel the airplane responding to his control movements. He managed to roll the aircraft to a wings-level position and, although still traveling at an extremely high speed, was finally able to pull the aircraft out of the dive—at the 5,000 foot level.

With the wings still in danger of being torn off, the Captain's next priority was to bleed off the excess speed. He pitched the nose of the airplane upward into a steep climb using the moon as a visual reference to maneuver the aircraft. Now the airspeed slowed rapidly and, with the assistance of the other two flight crew members, he was able to reduce it down to normal cruise, and level off at 13,000 feet.

Let us examine just how fast this airplane was traveling in that nose-dive towards the earth.

The descent of a Boeing 727 airliner, for the comfort and safety of its passengers, is approximately 1,500 feet a minute. Therefore, a 34,000-foot descent, in normal flight, would take approximately 23 minutes of flying time.

Flight 841 made that descent from 39,000 feet to 5,000 feet in only 63 seconds! Sixty-three seconds compared to 23 minutes. And, to show you how close to oblivion this aircraft was, only nine more seconds of diving at this incredible speed, would have sent Flight 841 rocketing straight into the ground.

Now, back to this resurrected flight. After regaining control, the flight crew noticed a warning light indicating a failure of one of the flight systems. Therefore, the Captain decided not to continue to Minneapolis but to land the aircraft at the nearby Metropolitan Airport in Detroit, Michigan. He instructed the First Officer and flight engineer to perform emergency checklist procedures and to notify the flight attendants to prepare the passengers for an emergency landing.

When the landing flaps were extended during the approach, the aircraft rolled sharply to the left. The Captain, therefore, retracted the flaps and planned for a landing without them.

However, there was another problem—the main landing gear indicators showed unsafe landing gear conditions. The Captain decided to fly down the runway at a low altitude to allow the control tower to make a visual check of the landing gear. The report he received from the tower and crash rescue personnel was that all three landing gears appeared to be fully extended. Hearing this, the Captain made his turn back to the runway and, as the passengers collectively held their breath, landed the aircraft safely at 10:31 PM.

Since the crew members were not sure how badly damaged their aircraft was, they decided not to risk taxiing to the terminal. They pulled into a turnoff from the runway, shut the engines down and radioed the tower for assistance to remove the passengers.

At this point, we would like you to spend a few minutes listening to a tape from the cockpit voice recorder revealing some of the conversations from within the aircraft. Although it does not give you a clue as to the reason for the erratic behavior of the aircraft, it does have relevance to some critical comments later made by the National Transportation Safety Board concerning the conduct of the crew.

| INTRA-COCKPIT | AIR-GROUND COMMUNICATIONS | |
|---|---|---|
| | Fire Dept: | Hello cockpit. |
| | Capt: | Yeah. |
| | Fire Dept: | Did you call operations and request a bus? |
| | Capt: | No, but we will. |
| | Fire Dept: | Okay, thank you. |
| | Capt: | Ah, ramp TWA, this is eight forty-one. |
| | TWA Ramp: | Yeah, go ahead. |

Capt:      We've been asked
           to deplane the
           passengers, ah,
           because of a slight
           fuel leak here . . .
           the fire depart-
           ment has asked us
           to get 'em off and
           we'd like some
           kind of transpor-
           tation, a bus for
           them, please . . .
           what we're going
           to do is drop the
           aft stairs and let
           them walk off
           without excite-
           ment, we just
           want to get them
           off easily, but we
           need to get them
           out or off the taxi-
           way here.

TWA Ramp:  Yeah, are you
           still, you still on
           the runway?

Capt:      No, we're on a
           turnoff from the
           runway, we're
           clear of the run-
           way.

TWA Ramp:  Okay, we'll see
           what we can do
           here, is there any
           way that you can
           keep in contact
           with us here?

|  |  |
|---|---|
| Capt: | Ah, I'm talking to you from the airplane right now. |
| TWA Ramp: | I mean you can stay on this frequency though? |
| Capt: | Yes I can. |

UNKNOWN VOICE: Did you feel kind [of] helpless in that seat back there.

UNKNOWN: Well, I'll tell you . . . believe me, definitely.

UNKNOWN: You know it's funny to be back here trying to analyze—this situation.

UNKNOWN: Hard to see what's happening, you guys were trying to pull it up.

UNKNOWN: Yeah.

UNKNOWN: Saying get it up, pull it up.

UNKNOWN: That's ah [deleted words] emergency descent, as a flyer who wasn't flying it, did all right, well done.
[Sound of cough]

UNKNOWN: What are you eating, you got one
of those cough
drops?

UNKNOWN: I'll get you one.

UNKNOWN: Throat a little dry.

UNKNOWN: Yeah a little dry
and my mouth's a
little dry.

Flt. Eng: Okay, I'll stay
here to stay on the
radio.

TWA Ramp: Eight forty-one
from Detroit
ramp.

Flt Eng: TWA's eight
forty-one, go
ahead.

TWA Ramp: Yes sir, looks like
you're pretty close
to Eastern's termi-
nal there, you
think its conceiv-
able that we can
walk the people
over there? I'm
gonna have a hard
time gettin' a bus.

Flt. Eng: Okay, if you
could bring some-
body over as a
guide, I think that
would be fine,
they wouldn't
mind walking that
far.

TWA Ramp: We'll do that.

| | |
|---|---|
| Flt. Eng: | Okay, what they intend to do is, they cannot get a bus so they're going to bring a guide out and walk them to the Eastern terminal. |
| UNKNOWN: | They won't leave them on the airplane. |
| Flt. Eng: | No, I don't imagine they will now. |
| Capt: | For all the help the people did great, they did exactly what they were told to do. |
| Flt Eng: | That's because you guys took over and did it. |
| Capt: | There were times on there when I had problems just looking to see if it was over with. |

| | |
|---|---|
| Flt. Eng: | Ah, ramp, TWA's eight forty-one. |
| TWA Ramp: | Go ahead. |
| Flt. Eng: | Do you need any further contact here, if not I'll turn the radios off. |

TWA Ramp: Ah, no except, ah, can you give me anything, any indication on the airplane or anything, everybody else is calling, is there any information that you can give me.

Flt. Eng.: No sir, we can't, I'm sitting in the cockpit and I can't tell you, I don't know what the situation is, you'll have to talk to maintenance.

TWA Ramp: Yeah, well, I mean you lost hydraulic, is that it?

Flt. Eng: We assume that's what happened but we can't tell . . . talk to maintenance.

TWA Ramp: Okay, you can sign off then.

Flt. Eng: Detroit ramp, do yo read [hear me]?

(End of recording)

After the passengers were safely deplaned, the aircraft was carefully examined. It was found to be substantially damaged from its acrobatic maneuvers at extremely high speed. The Number 7 leading edge slat on the right wing was missing.

(Remember this fact. It will become a key issue in the investigation.) There was also damage to the ailerons, flaps, main gear and nose gear. In addition, wing and fuselage skin panels were buckled or wrinkled, a hydraulic line ruptured and fuel was leaking from the left wing.

Despite all of these damages as well as numerous cylinders, rods and pistons found to be broken, bent, or missing, the aircraft was able to be repaired and returned to service in about six weeks.

As for the crew and passengers on Flight 841, they really had a miraculous escape from serious injury. Only eight of the 82 passengers suffered some form of minor injury.

Of the five passengers who initially reported injuries, two were taken by ambulance to a local hospital where they were treated and released. Three passengers reported pains in their chests, necks and backs, but refused medical treatment.

All five injured passengers were placed on another flight the next day and flown to their Minneapolis-St. Paul destination. It wasn't until later that three additional passengers reported that they were injured. One of them was hospitalized for severe muscle strain of the back and neck and a vertigo-balance problem.

Airline officials also arranged for the uninjured passengers to be flown to Minneapolis aboard another airliner. (However, one wary passenger decided to continue his trip by rented automobile.) Vouchers for food and drinks were handed out while arrangements were being made. Said one of the survivors, "I doubt there were too many who went for food."

The flight data recorder on the aircraft indicated that the crew members and passengers were subjected to an extremely high in-flight load factor during the dive from 39,000 feet and the subsequent recovery from 5,000 to 13,000 feet. This was as high as six G's, a gravitational weight of six times the gravity force we are normally subjected to. This was not only unusually high in force, but in duration, as well.

"You couldn't move your arms," one passenger recalled. "It was like you were glued in your seat. I was just plain frightened and horrified."

The high G's forced the occupants' heads and upper extremities toward the floor of the cabin and caused muscle strains of the neck and back. Passengers who were standing when the maneuver began, were forced to the floor and, in the process, contacted objects that also caused bruises and cuts.

One passenger was putting on fresh makeup in the restroom when she thought the airplane hit turbulence. Before she could gather up her cosmetics, she was thrown to the floor.

"I wasn't able to stand up. I couldn't breathe. I felt like I was being smashed," she recalled.

She said her arm, leg, hip and tailbone were bruised from her having been thrown against the toilet. She felt that she would not have been injured if she had been in her seat at the time.

Two of the passengers were a college student and his wife, who was six months pregnant. They were seated near the kitchen area of the plane on the way back from a trip to Spain.

"At first, I thought the pilot was trying to avoid another plane," said the student. "Then the plane just kept going faster, people began to scream.

"I looked around me and there was a stewardess sitting right behind me, crying. I thought about my family and about my wife and my little baby I thought I'd never get to see."

He said that after the plane leveled off, the pilot seemed pretty calm and told the passengers over the intercom, "There seems to be a small problem but we have it under control."

The flight attendants then gathered near the plane's kitchen, "closed the curtain and had a meeting," he said, "and began to throw pillows, coats and other things in front of us. They told us to take off our shoes, remove sharp objects from our pockets and get ready for evacuation.

"We passed over the airport one time. The Captain told us it was so people on the ground could look at the damage. It was awful tense. You could see blinking lights and fire trucks all over the runway," he said. "Then we landed and it was much smoother than we expected."

It is fascinating how many passengers turned to thoughts of love or affection at this moment of crisis. One man turned to kiss his wife. Another remembered that he hadn't told his wife in a morning telephone call that he loved her. Another woman said to her husband, "Well, we've had a good life." Then they kissed each other, "and that was it." Another woman had, "crazy thoughts like, did I kiss my husband good-bye?" She had kissed him, but not her child who dislikes partings. "I know I always will in the future, that's for sure," she said.

The Safety Board was notified of the accident the following morning and sent investigators to Detroit. In addition, representatives from the Federal Aviation Administration, Trans World Airlines, Inc., Boeing Company, and the Air Lines Pilots Association were asked to participate in the investigation.

One of the facts to emerge was that the Captain of Flight 841 was an experienced aerobatic pilot. Very few airline pilots have had this type of training or experience since it has never been deemed necessary to develop this skill in order to captain a commercial passenger aircraft. This is one case where the participants, in what turned out to be a wild, tumbling ride through the skies, can consider themselves fortunate in having this pilot at the controls. Many air observers believe that not many pilots would have been able, as this Captain did, to safely bring the aircraft to a successful landing after entering the vertical dive at 39,000 feet.

A Federal Aviation official praised the pilot for his skill and ingenuity. "There is nothing in the manual to tell you what to do in such a situation," he said. After another moment, he added, "They also had another pilot, and that was the good Lord."

Aviation authorities stated that it was a "miracle" the TWA jet survived and that none of the 89 people aboard were killed. "I can't think of any other incident where a commercial, passenger-carrying plane had done a complete 360 degree rollover and survived," said an official of the Federal Aviation Administration. "The miracle is that it held together under such extraordinary speed and circumstances." He added, "Obviously, this airframe (Boeing 727) is strong as a brick."

The three-engine 727 is one of the most common airplanes in commercial use. The Boeing Company says it has delivered more than 1,400 of the planes to 89 airlines around the world, and one lands or takes off every seven seconds around the clock. To date, these planes have carried more than a billion passengers and about 12 million persons are flown in a 727 every month.

At the request of the Safety Board, the Boeing Company programmed a flight simulator with Flight 841's gross weight and center of gravity conditions and the pertinent data associated with its flight. A total of 118 simulated flights were conducted in the simulator to identify the condition that precipitated the aircraft's upset and to duplicate and evaluate its maneuver.

Also at the request of the Safety Board, the aircraft manufacturer conducted flight tests in an instrumented Boeing 727. The purpose of the tests was to record data that could be compared with Flight 841's flight data recorder.

After an extensive investigation, the National Transportation Safety Board came to a number of findings and conclusions in regard to this accident:

1.  The 34,000 foot dive of the TWA 727 was caused by an extended right wing slat that could not be retracted. (Slats are lift enhancement devices located on the leading edges of the wings.)

2.  Contributing to the cause was a pre-existing misalignment of the Number 7 slat which, when combined with air pressure upon it at cruise speed, prevented successful retraction. (There are eight slats on a Boeing 727, numbered in order from the pilot's left.)

3.  The slats—or the Number 7 slat alone—could not have been extended by mechanical malfunction. Although flight crew members testified that they did not extend the slats, the Safety Board could not accept their statements. From an analysis of the results of both flight simulation and actual flight testing, the Board found that "the possibility of a series of malfunctions and failures occurring which permitted the slat to extend aerodynamically or hydraulically is extremely remote."

4.  Comparison of flight test data with the flight recorder data from the accident airplane shows that the Numbers 2, 3, 6 and 7 slats were extended "as a consequence of flight crew action." Those slats were then retracted with the exception of the Number 7 slat, which remained extended because of the misalignment.

5.  Although it cannot determine why the four slats were extended, the Board listed as a possibility "an effort to improve aircraft performance [speed]."

6.  Although the Captain's extension of the landing gear in an

effort to regain control significantly reduced the speed of the dive, "recovery would have been doubtful" if the one extended slat (Number 7) had remained in place. The Board said the 727 remained uncontrollable until the Number 7 slat tore loose from the wing when the airliner dove below 15,000 feet.

One other intriguing aspect of this accident was the fact that 21 minutes of the 30-minute tape taken from the cockpit voice recorder turned out to be blank. This was a source of disappointment to the investigating group. After any accident, listening to the tape of the cockpit voice recorder is of extreme importance to investigators to pick up pertinent information as to flight crew performance.

What was especially frustrating was the fact that the remaining nine minutes of the tape were of good fidelity but pertained only to flight crew conversation *after* the aircraft was on the ground at Detroit.

Tests of the cockpit voice recorder (CVR) in the aircraft showed that it was not damaged. The CVR tape can be erased by means of the bulk-erase feature on the CVR control panel located in the cockpit. This feature can be activated only after the aircraft is on the ground with its parking brake engaged.

In a deposition taken by the Safety Board, the Captain stated that he usually activates the bulk-erase feature on the CVR at the conclusion of each flight to preclude inappropriate use of recorded conversations. However, in this instance, he could not recall having done so. The First and Second Officers both stated that they did not erase the tape nor did they see the Captain activate the erase button on the CVR control panel.

The National Transportation Safety Board responded to this testimony by making this statement:

"Since our weighing of the evidence involves a rejection of the possibility of an unscheduled extension of the No. 7 slat and a partial rejection of the captain's recollection of his actions following extension of the slats, the Safety Board believes that the following comments are appropriate: We believe the Captain's erasure of the CVR is a factor we cannot ignore and cannot sanction. Although we recognize that habits can cause actions not desired or intended by the actor, we have difficulty accepting the fact that the Captain's putative habit of routinely erasing the CVR after each flight was not restrainable after a flight in which disaster was only narrowly averted. Our skepticism persists even though the CVR would not have contained any contemporaneous information about the events that immediately preceded the loss of control because we believe it probable that the 25 minutes or more of recording which preceded the landing at Detroit could have provided clues about causal factors and might have served to refresh the flight crew's memories about the whole matter."

Obviously, there was a difference of opinion between the Safety Board and the Captain as to whether or not the CVR tape had been deliberately erased.

However, what cannot be erased or doubted is the fact that, with 89 lives hanging in the balance, this pilot, in that 63-second plunge from 39,000 feet, had to be credited with a professional and skillful performance in ultimately bringing the aircraft to, what turned out to be, a routine landing.

Chapter 16

# Strayed Off Course—Attacked By Fighter Aircraft

If wild acrobatic maneuvers are about the last thing that passengers would expect to encounter on a routine airplane flight, what can we say for those passengers flying high above the ocean when their helpless, unarmed airliner becomes the prey and target of unseen jet fighter aircraft? This most unlikely scenario actually took place.

A Korean 747 commercial jet aircraft with 269 persons aboard departed J.F. Kennedy Airport, New York, shortly after midnight on August 31, 1983. Among the passengers was a U.S. Congressman and 60 other Americans.

After making a refueling stop at Anchorage, Alaska, the airliner, Flight 007, resumed its tedious flight (total time, almost 15 hours) bound for its ultimate destination, Seoul, Korea.

The flight was uneventful until about one AM, September 1, Korean time, when the 747 jet, programmed to fly just south of the Soviet Union boundary, started to stray off course and drifted north into Russian airspace. This immediately resulted in a number of Soviet supersonic jet fighters scrambling to track the "alien" aircraft flying at 33,000 feet. As the airliner with its 269 civilian occupants flew over sensitive Soviet missile-testing facilities, the tiny Russian jets swarmed under and around the jumbo aircraft, like gnats, for approximately 2-and-a-half hours, seeking instructions from their ground command.

It was now reaching the point where a decision would have to be made. The speedy fighter planes would soon be running out of fuel and the airliner would shortly be flying out of the Soviet airspace into international territory over the Sea of Japan. To add to the drama at the moment of truth, there is no evidence that any communication had been established between the stalkers and their quarry or that the crew was even aware of the fact that they were off course and pursued by fighter airplanes.

We now turn to a transcript of tapes furnished by Japanese intelligence stations of actual transmissions by the Soviets. The following is an edited account (for the sake of brevity) of the jet fighters communicating with each other, as well as their responses to orders from their ground command:

Fighter 1:    I didn't understand. What course? My course is 100 [degrees].

Fighter 2:    I see it. Roger, understood. I'm flying behind.

Fighter 3:    Executing course 100.

Fighter 3:    [Jet fighter 3] needs to drop his wing tanks.

Fighter 2:    Yes. It has turned. The target is 80 degrees to my left.

Fighter 1:    Executing [Responding to order from ground command].

Fighter 2:    Executing.

Fighter 3:    Executing.

Fighter 2:    I see it visually and on radar.

Fighter 3:    I have dropped my tanks. I dropped them. Executing.

Fighter 2:    I see it. I'm locked on to the target . . . The pilot isn't responding to IFF [Identify, Friend or Foe]. . . . [The weapons system] is turned on.

| | |
|---|---|
| Fighter 2: | Roger. I have [enough] speed. I don't need to turn on my afterburner. . . . My fuel remainder is 2700. |
| Fighter 3: | I've dropped my tanks, one at 4000 [feet], one at 3000. |
| Fighter 2: | The target's course is still the same. 240 [degrees]. |
| Fighter 3: | Executing. |
| Fighter 2: | I am in lock-on. |
| Fighter 3: | [Ground command] is inquiring: Do you see the target or not? |
| Fighter 2: | I see it. . . . The A.N.O. [air navigation lights of the airliner] are burning. The [strobe] light is flashing. |
| Fighter 3: | Roger. I'm at 7500 [feet]. Course 230 [degrees]. |
| Fighter 2: | I am closing on the target. |
| Fighter 3: | I am flying behind the target at a distance of 25. Do you see [me]? |
| Fighter 2: | Fiddlesticks! I'm going. That is, my Z.O. [indicator] is lit [missile warhead already locked on]. . . . The target's [strobe] light is blinking. I have already approached the target to a distance of about two kilometers . . . what are instructions? |
| Fighter 2: | The target is decreasing speed . . . Now I have to fall back a bit from the target . . . I'm dropping back. Now I will try a rocket. |
| Fighter 3: | 12 [kilometers] to the target. I see both. |
| Fighter 1: | I'm in a right turn on a course of 300 [degrees]. |
| Fighter 2: | I'm closing on the target, am in lock-on. Distance to target is 8 . . . I have executed the launch. |
| Fighter 2: | THE TARGET IS DESTROYED! |

It took approximately 12 agonizing minutes for the 747 jumbo jet to fall from 33,000 feet into the Sea of Japan. The shooting down of a civilian airliner with the loss of 269 lives from 13 different countries outraged the world of nations which found itself frustrated on how to appropriately respond.

However, nothing could compare to the shock, anguish and despair of the families of the victims left with nothing tangible to mourn. Perhaps, what could sum up their feelings, was the gesture of the father of one of the victims who travelled on a fishing vessel near where the airliner disappeared. He threw his son's sweater into the cold water and cried out, "My son, my son, you must be freezing."

Chapter 17

# Snow and Ice on the Wings

Up to now, you've read about many varied causes of aircraft accidents. Some of them have been due to only one unfortunate circumstance that arose; others were more complex, with a number of ill-fated conditions being blamed for the mishap.

We now come to the grand-daddy of them all. An accident in which so many factors played a part in causing this disaster, that we need more than all of the fingers of both hands to count them. It seems that almost any circumstance that could be responsible for contributing to the accident, existed at the time. Like Murphy's Law, "Whatever can go wrong, will—and, at the worst possible time."

Follow the sequence of events of this chapter and share our disbelief.

It was a cold, snowy day in Washington, D.C. on January 13, 1982. By late afternoon, the temperature was down to 24 degrees Fahrenheit, the winds were blowing at about 12 miles per hour, the clouds were as low as 200 feet above the ground and the visibility was reduced to less than a mile. Not exactly excellent flying conditions at the Washington National Airport.

Scheduled to fly that afternoon from Washington to Tampa and then to Fort Lauderdale, Florida, was Air Florida's Flight 90, a 737 jet airliner with 74 passengers, including 3 infants. The crewmembers consisted of a captain, first officer and 3 flight attendants. The scheduled departure time of 2:15 PM could not be kept. There were significant delays in all flights that day due to the heavy snowfall which resulted in the temporary closing of the airport.

Plows were initially used to remove the snow from the runway followed by brooms which swept away the loose snow after the plows had passed. The runway surface was then sanded. This procedure took an hour and 20 minutes before the runway could be reopened for traffic. Meanwhile, the snow continued to fall.

The stage is now being set for disaster due to the existence of an unbelievably large number of unfortunate circumstances.

Normally, as many as two or three coincidental conditions that would contribute towards an aircraft accident, would be considered to be unusual. What would you think if you learned that, in the chapter you are now reading, this particular tragic aircraft crash was due to 12 different circumstances, all playing a part in the destruction of the airplane?

Impossible you say? Judge for yourself as we review them in chronological order.

No. 1. The inexperience of the flight crew. The captain of Flight 90 was 34 years old, having served as a first officer for only 2 years before being upgraded to his present position with Air Florida in August 1980. (The Safety Board later stated that a survey of major trunk carriers showed that pilots upgrading to captain generally had served an average of 14 years as a first or second officer. The Board believed that this longer seasoning experience was missed in this case due to the recent rapid expansion of Air Florida wherein pilots were upgrading faster than the industry norm to meet the demands of growing schedules.) A review of the captain's operating experience also disclosed only 8 previous occasions where arrivals and departures were conducted during weather conditions conducive to icing.

As for the first officer, age 31, before employment with Air Florida, his experience was gained as a military jet fighter pilot.

He had been a first officer for 14 months and, in his case, there were only 2 occasions during that period where he had conducted ground operations in conditions conducive to icing.

Thus, according to the Safety Board's report, "neither of the flight crew had much experience in operating jet transport aircraft in weather conditions like those in Washington National Airport on January 13, 1982."

No. 2.  Inadequate runway facilities. Although Washington National Airport is the recipient of a heavy air traffic flow, it has always had problems in icing conditions. The airport only has one instrument runway and therefore, it is necessary to close the airport when snow removal is required. (Most other major airports have multiple runways enabling them to continue operations while one runway is being plowed and sanded.) Closing the only operating runway for snow removal creates significant air traffic delays with a backlog of departing aircraft. This situation creates pressure on flight crews anxious to maintain schedules.

No. 3.  Inadequate ground facilities. Washington National gate and ramp space is limited. When traffic is very heavy, it is necessary to instruct aircraft ready for departure to leave the gate and taxi behind a long lineup of airplanes in order to provide space for arriving traffic.

This presents a special problem for aircraft that have to be deiced. Deicing is absolutely necessary because even a small amount of snow, slush or ice adhering to the surface of an aircraft, particularly the wings, will have a significant negative effect on the performance of an aircraft. It is the smooth flow of air over the wing which produces the lift to support the weight of the aircraft. Snow and ice contamination on the wing will have the effect of changing the contour shape and reduce the wing's lift-producing efficiency. When the wing's efficiency is reduced to the point where even higher airspeed cannot compensate for the loss of lift, a stall will occur.

The Air Florida Maintenance Manual specifically states, under the heading of Cold Weather Procedures, that "No aircraft will be dispatched and no take-off will be made when the wings, and/or tail surfaces have a coating of ice, snow or frost."

Deicing normally takes place at the gate. However, after

deicing, if an aircraft is forced to give up its gate position and taxi behind other aircraft awaiting takeoff, it is possible that snow and/or ice will re-form before reaching the position of number one for takeoff. In that case, the airplane will have to surrender its precious place in line and taxi back to the gate for additional deicing—if space is still available there. And, after deicing, take its place, once again, at the end of the line.

Obviously, this can influence the judgment of an impatient flightcrew who have already been subjected to a long delay. (Flight 90 had been informed that there were 16 aircraft that had departure priority before it could even push back from the gate to start its taxi.)

No. 4.  A peculiarity of the 737 airplane. There is an additional problem of snow or ice on the wings of an aircraft which specifically applies to the Boeing 737 jet. Since 1970, there have been 22 reports made by operators of 737 aircraft that when ice or frost was present on the wing leading edges, this model aircraft experienced a sudden pitchup or rolloff immediately after takeoff. In a number of these incidents, the aircraft's stall warning system activated and it was necessary for the pilot to use full control movements to recover. This sensitivity led to the Boeing Company issuing a Bulletin cautioning 737 flightcrews to comply with all ice and snow removal procedures prior to take-off.

We now get back to Flight 90, at its gate, requesting deicing. This started at 2:45 PM and continued until 3:00 PM, completing the left side of the aircraft. The operator of the deicing vehicle was then replaced by a relief operator who deiced the right side of the airplane and completed the task at 3:10 PM.

The special deicing vehicle uses a combination of hot water and glycol sprayed on the aircraft with a high-pressure hose. Deicing solution manufacturers recommend that aircraft surfaces be sprayed with a high glycol solution if conditions are conducive to refreezing after deicing. With a high concentration, the solution can remain on the surfaces of the aircraft and provide extra protection against the formation of ice.

No. 5.  Incorrect deicing fluid mixture. In the case of Flight 90, a mixture of 70% water and 30% gylcol was set by the operators as the proper mix for that aircraft. Unfortunately, the standard nozzle on the deicing hose which accurately calibrates and visually provides the operator with a reading of the mix, was replaced with another that did not contain these features. Tests of the solution taken from the deicing vehicle afterwards indicated that, instead of a 70% water-30% deicer fluid mixture, the aircraft received a mixture of 82% water and only 18% deicer fluid. Evidence provided by a photograph taken by a passenger on another flight only 10 to 15 minutes after the completion of Flight 90's deicing, indicated that new snow had already accumulated on the top and upper right side of the aircraft's fuselage.

No. 6.   Lack of follow-through by the crew. When the deicing operations were completed, the captain, who was sitting in the left cockpit seat, asked the Air Florida station manager, who was standing near the main cabin door, how much snow was on the aircraft. The manager reported there was a light dusting of snow on the left wing from the engine to the wingtip but that the area from the engine to the fuselage was clear. (The Safety Board's report pointedly stated, "Although the captain was solely responsible for assuring that the aircraft was ready for flight, no witnesses recalled seeing either the captain or the first officer leave the cockpit to inspect the aircraft from outside for remaining snow or ice contamination . . . Good practice dictates that one of the flight crewmembers observe the aircraft from the outside."

At 3:16 PM, the following radio transmissions took place:

Flight 90:     [To] Ground [control], Palm 90 [air traffic control designation for Air Florida Flight 90] like to get in sequence, we're ready.
Ground control: Are you ready to push?
Flight 90:     Affirmative.
Ground control: Okay, Palm Ninety, push approved.

No. 7.   Improper use of engines. At 3:25 PM, a tug attempted to push Flight 90 back from the gate so it could start to taxi. However, the ground vehicle, which was not equipped with chains, could not move the aircraft. When a flight crewmember suggested to the tug operator that the aircraft's engine reverse thrust be used to push the aircraft back, the operator stated that this was contrary to the policy of the airline. Nevertheless, the aircraft's engines were started and operated in reverse thrust for a period of 30 to 90 seconds. During this time, several airline personnel observed slush being blown toward the front of the aircraft. (Air Florida Manuals disseminated to all their flightcrews, contains this warning, "A buildup of ice may occur during ground operations involving use of reversers in light snow conditions. Snow is melted by the deflected engine gases and may refreeze as clear ice upon contact with cold leading edge devices.")

Finally, with the help of another tug, Flight 90 was freed from the snow and ice and pushed back from the gate. At 3:35 PM, the engines were restarted and the aircraft was ready to taxi.

| | |
|---|---|
| Ground Control: | [To another aircraft] Can you get around that Palm [Air Florida Flight 90], on the pushback? |
| Flight 90: | Ground, Palm 90, we're ready to taxi out of his way. |
| Ground Control: | Okay, Palm 90, roger, just pull up over behind that TWA and hold right there, you'll be falling in line behind a Apple [New York Air] DC-9. |
| Flight 90: | [Acknowledges] Palm 90. |
| 1st. Off. (intra-ckpt): | [To captain] Behind that Apple, I guess. |
| Capt. (intra-C): | Behind what TWA? |

(All conversations now intra-cockpit except where indicated otherwise.)

| | |
|---|---|
| 1st Off.: | Over by that TWA to follow that Apple, apparently. |
| 1st Off.: | Boy, this is shitty, it's probably the shittiest snow I've seen. |
| 1st Off.: | It's been awhile since we've been deiced. |
| Capt.: | Think I'll go home and. . . . [unintelligible]. |
| 1st Off.: | That Citation [small private jet] over there, that guy's about ankle deep in it. |
| | [Sound of laughter] |
| 1st Off.: | Hello Donna. |
| Head Stewardess: | I love it out here. |
| 1st Off.: | It's fun. |

| Head Stewardess: | I love it. |
| 1st Off.: | See that Citation over there, looks like he's up to his knees. |
| Another stewardess: | Look at all the tire tracks in the snow. |

While this conversation was taking place, the long line of aircraft slowly kept moving up towards the take-off position. The captain now intentionally taxied his airplane close behind the aircraft in front of him in an attempt to use the heat and blast from the exhaust gases of that aircraft to remove snow and slush from his wings. That is No. 8 in the number of circumstances. The Safety Board called specific attention to the cold weather procedures of the FAA Flight Manual issued by the Boeing Co. which states, "Maintain a greater distance than normal between airplanes when taxiing on ice or snow covered areas. Engine exhaust may form ice . . . and blow snow and slush which freezes on surfaces it contacts."

| Capt.: | Tell you what, my windshield will be deiced [from the exhaust of the airplane ahead] don't know about my wing. |
| 1st Off.: | Well, all we really need is the inside of the wing anyway, the wing tips are gonna speed up by eighty [knots] anyway, they'll shuck all that other stuff. |
| Capt.: | [Gonna] get your wing now. |
| 1st Off.: | D'thing they get yours? Can you see your wing tip over 'er? |
| Capt.: | I got a little on mine. |
| 1st Off.: | A little . . . this one's got about a quarter to half an inch [of snow] on it all the way. |
| 1st Off.: | [Speaking about another aircraft that just landed] Look how that ice is just hanging on his back, back there, see that? |
| 1st Off.: | It's impressive that these big old planes get in here with the weather this bad, you know, it's impressive. |

| | |
|---|---|
| 1st Off.: | It never ceases to amaze me when we break out of the clouds, there's the runway anyway, don't care how many times we do it. God, we did good [sound of laughter]. |
| 1st Off.: | I'm certainly glad there's people taxiing on the same place I want to go cause I can't see the runway [and] taxiway without these flags. |
| Other Stewardess: | We still fourth [to take off]? |
| 1st Off.: | Yeah. |
| Other Stewardess: | Fourth now. |
| 1st Off.: | We're getting there, we used to be seventh. |
| Capt.: | Don't do that Apple, [evidently pulling ahead]. I need to get the other wing done [sound of laughter]. |
| 1st Off.: | Boy, this is a losing battle here on trying to deice those things, it [gives] you a false feeling of security, that's all it does. |
| Capt.: | That satisfies the Feds [Federal Aviation authorities]. |
| 1st Off.: | Yeah. |
| Capt. | Right there is where the icing truck [should be] they oughta have two of them, you pull right. . . . |
| 1st Off.: | Right out. |
| Capt.: | Like cattle, like cows, right in between those things and then . . . |
| 1st Off.: | Get your position back. |
| Capt.: | Now you're cleared for takeoff. |
| 1st Off.: | Yeah, and you taxi through kinda like a car wash or something. |
| Capt.: | Yeah, hit that thing with about eight billion gallons of glycol. |
| Capt.: | In Minneapolis, the truck they were deicing us with, the heater didn't work on it, the [deleted word] glycol was freezing the moment it hit. |

1st Off.:          Boy, I'll bet the school kids are just [deleted word] in their pants here. It's fun for them, no school tomorrow, yahoo [sound of laughter].

As Flight 90 slowly taxied until it was almost next in line to assume the takeoff position on the runway, both pilots went through the normal cockpit procedure of using a check list to remind them that the aircraft doors were closed, flaps set, controls operating freely, instruments set properly, etc., etc. Here we come to one of the most important circumstances, No. 9, failure to turn on the engine anti-ice system.

On the checklist was the item of anti-ice. The function of the anti-ice system is to maintain a flow of heated air, in icing conditions, to a probe that determines engine pressure. Without this heat, ice is apt to accumulate, blocking the inlet of the probe which will give a false reading of takeoff thrust on the engines.

Despite the flightcrew's use of the checklist as a reminder, the engine anti-ice system was left in the "off" position. When the probe is blocked by ice, the engine pressure reading (EPR) will indicate a much higher thrust than actually exists. Since the EPR reading called for a setting of 2.04 on takeoff, because of the blocked probe, the actual thrust generated by the engines will be the equivalent of 1.70 with an unblocked probe—considerably less takeoff thrust than normal.

1st Off.:     Slushy runway, do you want me to do anything special for this [takeoff] or just go for it.
Capt.:        Unless you got anything special you'd like to do.
1st Off.:     Unless just take off the nose wheel early, like a soft field takeoff or something.
1st Off.:     I'll take the nose wheel off and then we'll let it fly off.
Tower:        [To] Palm 90, taxi into position and hold, be ready for an immediate [takeoff].
1st Off.:     [To tower, acknowledges] Palm 90, position and hold.

This last recording was followed by sounds picked up on the

cockpit microphone of the parking brake being let off, the takeoff warning signal alerting the flight attendants and the flap levers being activated. The next transmission was that coming from the public address system, "Ladies and gentlemen, we have just been cleared on the runway for takeoff, flight attendants, please be seated."

| | |
|---|---|
| Tower: | Palm 90, cleared for takeoff. |
| 1st Off.: | [Acknowledges] Palm 90 cleared for takeoff. |
| Tower: | No delay on departure if you will, traffic's two and a half out for the runway [another aircraft is two and a half miles away from landing on the same runway]. |
| 1st Off.: | Okay. |
| Capt.: | Your throttles [the captain delegates the handling of the controls on takeoff to the 1st officer]. |
| 1st Off.: | Okay. |

It is now all of 50 minutes since the aircraft had finished being deiced. Since the Boeing 737 had been exposed to moderate to heavy snowfall during that long taxiing period, the Safety Board feels that, "the evidence is conclusive that the flightcrew was aware that the top of the wings were covered with snow or slush before they attempted to takeoff." And, "probably influenced by the prolonged departure delay . . . was thus hesitant to forego the takeoff opportunity and return to the ramp for another cycle of deicing and takeoff delay."

A captain of another aircraft, taxiing along the runway, witnessed Flight 90's takeoff roll and discussed the extensive amount of snow on the fuselage. "I commented to my crew, 'look at the junk on that airplane,' . . . Almost the entire length of the fuselage had a mottled area of snow and what appeared to be ice . . . along the top and upper side of the fuselage above the passenger cabin windows. . . ."

That was circumstance No. 10. The Federal Aviation Regulations are very specific, requiring that "no person may take off an aircraft when frost, snow or ice is adhering to the wings, (or) control surfaces . . . of the aircraft."

It is now just 12 seconds short of four PM when the throttles are pushed forward for the takeoff. The following transmissions are all intra-cockpit:

| | |
|---|---|
| Cockpit Recorder: | Sound of engine spoolup [power being applied]. |
| Capt.: | Holler if you need the wipers. |
| Capt. | It's spooled. |
| Unknown voice: | Whoo. |
| Capt.: | Really cold here [evidently some dials on the panel indicating lower than expected]. |
| 1st Off.: | Got 'em? |
| Capt.: | Real cold. |
| 1st Off.: | God, look at that thing [the instruments on the panel]. |
| 1st Off.: | That don't seem right, does it? |
| 1st Off.: | Ah, that's not right. |
| Capt.: | Yes it is, there's eighty [knots—the first check point]. |
| 1st Off.: | Naw, I don't think that's right. . . . Ah, maybe it is. |
| Capt: | Hundred and twenty [knots—the second check point]. |
| 1st Off.: | I don't know [still doubtful]. |
| Capt.: | V-1 [known as the decision speed. Up to V-1, the captain can safely abort the takeoff. Since the speed of V-1 had already been reached, the aircraft was now committed to takeoff]. |

No. 11.   The captain's failure to take action. In the conditions existing that day, a 737 jet should accelerate in about 30 seconds to a liftoff using 3,500 feet of runway. This airplane took 45 seconds and 5,200 feet of runway to liftoff—which was 50% longer in both time and distance.

It was apparent to the 1st officer that the aircraft was not accelerating properly. He expressed his doubts and concerns as many as 5 times in a period of 25 seconds during the takeoff roll. The captain chose to ignore the comments of his 1st officer and continued the takeoff.

The Safety Board stated that, "An observation that something is not right is sufficient reason to reject a takeoff without further analysis. The problem can then be analyzed before a second takeoff attempt." The Air Florida Flight Operations Manual emphasizes, "No matter which crewmember is making the takeoff, the captain ALONE makes the decision to 'REJECT.'"

Flight 90, past V-1, continues accelerating on its takeoff roll.

| | |
|---|---|
| Capt.: | Easy. |
| Capt.: | V-2 [after rotating off the runway, the V-2 climb speed of 165 miles per hour is reached]. |
| Cockpit Recorder: | [Sound of stickshaker starts—which is a device to warn the crew of an impending stall]. |
| Capt.: | Forward, forward [evidently the aircraft nose is pitching up too sharply]. |
| Unknown voice: | Easy. |
| Capt.: | Come on, forward. |

On takeoff, a 737 aircraft should climb at about 2,000 feet a minute. The flight data recorder indicated that Flight 90 climbed at only the rate of 1,200 feet a minute—and only for a short while. The highest altitude the aircraft reached was 352 feet while the airspeed began to dissipate. Meanwhile, the stickshaker continued to sound its warning. The aircraft desperately needed more power.

Which brings us to unfortunate circumstance No. 12. Flightcrews have been trained to avoid using full thrust on engines because of the concern about exceeding engine limitations. If the training had not been so ingrained, the crew would have disregarded the false engine pressure reading of 2.04 (which only generated 75% of takeoff power), and pushed the throttles all the way in.

According to the Safety Board, "had full thrust been added immediately following the activation of the stickshaker, the aircraft could probably have accelerated to a safe stall margin and continued flying."

| | |
|---|---|
| Capt.: | Forward. |

| | |
|---|---|
| Capt.: | Just barely climb [evidently the nose of the aircraft is still pointing up too sharply— wants it lowered]. |
| Unknown Voice: | Stalling—we're falling. |
| 1st Off.: | Larry, we're going down, Larry. |
| Capt.: | I know it. |

Both pilots were now frantically trying to coax the aircraft into, at least, maintaining its altitude. However, despite their desperate efforts and their fervent urgings (with, perhaps, the use of body English), the airliner, reluctantly thrusting forward through the moist air, was sinking rapidly.

At 4:01:01 PM, just 73 seconds after the start of the takeoff roll, 103,000 pounds of aircraft descended (despite the nose pointing up) into the heavily congested northbound span of the 14th Street Bridge, which connects the District of Columbia with Arlington County, Virginia. The bridge was jammed with the cars of homeward-bound commuters who had left work early to try to avoid the worst of the snowstorm.

As it pancaked, the rear half of the jet aircraft demolished the tops of 6 automobiles and a 10-wheel boom truck, killing and injuring many of the occupants as well as mangling and destroy-ing the vehicles. It then struck the bridge itself and ripped away a large section of the bridge wall and railings as it caromed across the span. The impact of the tail section with the bridge jolted the front of the airplane down as it tore past the other side of the span. Now in a nose-down position, with the passengers still strapped in their seats, it plunged completely through the ice cov-ering the Potomac River, where it disappeared from view. Heavy snow continued to fall.

Ground witnesses generally agreed that the aircraft was flying at a nose-high attitude of 30–40 degrees before it hit the bridge. Four persons in a car on the bridge claimed that large sheets of ice (evidently from the airplane) fell on their automobile.

Said a driver whose car was on the bridge, "I heard screaming jet engines . . . The nose was up and the tail was down. It was like the pilot was trying to climb but the plane was sinking fast . . . I saw the tail of the plane tear across the top of the cars, smashing some tops and ripping off others . . . Once the tail was across the bridge, the plane seemed to continue sinking very fast . . . [and] . . . hit the water intact in a combination sinking/ plowing action. I saw the cockpit go under the ice. I got the impression it was skimming under the ice and water."

Another witness on the bridge said, "It went down [under the water] nose first. When it hit the ice, it sounded like a pane of glass breaking, like a big rock had been thrown through a giant piece of glass."

Visibility was so restricted that the local controller could only track the takeoff progress of Flight 90 on his radar monitor. Within 20 seconds of the aircraft's contact with the bridge, he noted that Flight 90 had disappeared from his radar screen. He immediately notified Operations and Safety and called the Washington National Airport's fire department to report a possible accident. Meanwhile, another controller continued attempts to establish communications with Flight 90 until the tower received a telephone call that an aircraft had crashed north of the 14th Street Bridge.

Within five minutes, all neighboring police and fire departments were notified and responded to the emergency. However, none were properly equipped to perform a rescue operation on the ice-coated river. The airport airboat had difficulty maneuvering on the ice and never reached the scene in time to rescue survivors. The fire boats and harbor police boats were unable to break the ice and, therefore, also could not arrive in time to be effective.

The U.S. Park Police, with a helicopter not equipped nor required for rescue operations, turned out to be the mainstay of the rescue effort. Eagle 1, a jet-powered helicopter, with a crew of two Park Police officers, arrived on the scene shortly after being notified.

In the meantime, of the 79 people on board Flight 90, only five passengers and one flight attendant were able to escape being entombed in the submerged jet airliner. They had all been seated in the aft cabin area and were thrown out of the aircraft when the rear section separated from the main fuselage after striking the bridge and plowing into the Potomac River. These six were now clinging to the wreckage of the tail section, the only part of the 737 jet still afloat.

These floundering survivors had now been helplessly immersed for 22 minutes in the freezing waters before the National Park Service helicopter arrived on the scene. By that time, because of the intense cold, all of them had lost most of the effective use of their hands and were in danger of losing consciousness as well. Up to the point of arrival of Eagle 1, icing conditions were such that the major rescue effort consisted of firemen on the shore of the river bank, 100 yards away, desperately tossing lifelines to them, without success. They were just too far away. The only assistance available to them from the shoreline were the shouts of encouragement from people on the bridge to "hang in there, hang in there."

At last, there was the helicopter. The pilot swooped down and hovered the aircraft over the 6 survivors while his crewmen dropped make-shift rescue aids which consisted of ropes with hoops and life rings. This was the beginning of a number of heroic actions on the part of both the rescuers and some of those being rescued.

Let us start with the flight attendant. Also struggling to stay alive in the freezing water, she displayed unusual unselfishness by inflating the only available lifevest and giving it to one of the more severely injured passengers. Other lifevests were floating in the area but the survivors were unable to retrieve them. (They later reported that they experienced extreme difficulty in opening the package containing the one lifevest which was retrieved. The plastic package which contained the lifevest was finally opened by chewing and tearing at it with their teeth.)

Then there is the helicopter crew. At the risk of their own lives, the two crewmen were operating their flying machine in hazardous weather conditions at an extremely low and dangerous altitude. After dropping their looped ropes and life rings to the survivors fortunate enough to be able to grab and hold with their hands and wrists, they proceeded to drag their precious cargo to the shore where eager hands were waiting. The rescued were then treated on the scene for hypothermia (low body temperature) and shock before being placed in ambulances and rushed to the hospital.

To accomplish one rescue, the pilot performed an extremely dangerous maneuver to save a woman who was unable to hold on to the rescue line. He dipped one landing skid into the water while his crewman climbed out of the helicopter, stood on the skid and physically pulled that passenger right out of the water and aboard the same skid he was standing on. Eagle 1 then completed the rescue by gently depositing the injured woman at the foot of a waiting ambulance.

There were other outstanding acts of heroism. Of the six struggling people in the water, one of them actually gave up his own life so that the other survivors could be saved. When the hovering helicopter first dropped the rescue line to him, he helped place it around one of the women in the water so that she could be dragged to safety. Every time the line was dropped to him, he kept passing it off to one of the others. After the other five clinging to the wreckage, three women and two men, were, one by one, successfully plucked out of the icy water, Eager 1 finally returned to pick him up. But, too late. He was gone. The helicopter crew kept circling and circling looking for him. He was nowhere to be seen.

In a later interview, one of the helicopter Park Police officers stated that he had never seen such courage, not even in Vietnam, where he flew combat missions. "He could have gone on the first trip," said the pilot. "We threw the ring to him first, but he passed it to somebody else. We went back five times and each time he kept passing the ring to someone else. He is the hero of this whole thing." The officer added, "You have to ask yourself the question: If you were in his situation, a hundred yards from shore and knowing that every minute you were closer to freezing to death, could you do it? I really don't think I could."

Said the other helicopter officer, "I cried when I did not see him. If I could have seen him under the water, I would have plunged in myself to try to pull him out, dead or alive."

Both officers have a vivid memory of the man's face and were anxious to, at least, know who he was. Unfortunately, although all the bodies of the victims were eventually recovered, he could never be positively identified. Even the President of the United

States was moved to say this about the courage of this unknown hero, "We don't know who he is because he gave his life [to save the others]."

The tragic crash set the stage for another act of courage. A 28-year-old government employee, a father of two, was standing on the river bank watching the helicopter rescue operation. He saw an injured passenger being dragged to safety. However, dazed and in shock, she was unable to maintain her hold on the rescue line and, to the horror of the viewers on shore, fell back into the water. He quickly tore off his coat and boots and dived into the frigid Potomac River.

He recalled later, "When she let go of that life ring, I knew she would drown if I didn't jump in. There was no other way they could have gotten to her." He said, "Her body just went limp and she couldn't get the rope . . . I think she passed out. Her eyes rolled back and she started to go under.

"Something told me to go in after her. I jerked off my coat and dove in." He added, matter-of-factly, "I just did what I had to do."

Since television cameras were on the scene, millions of TV viewers were able to watch, breathlessly, as he swam out to her, cradled her head above the water and fought his way inch by inch, through the ice towards the river bank where eager hands reached out for them. Both of them were wrapped in blankets, treated for hypothermia and rushed to the hospital.

This hero did not show up in his office the next day. "My supervisor at work saw it on TV and she said that I deserved the day off and that if I went to work, she would fire me," he laughed.

It is interesting that it took the heroic deeds of five individuals to save five survivors: Two operators of the rescue helicopter, one who sacrificed his own life, the flight attendant who gave up the lifevest and the government employee who jumped into the icy waters. Five for five—with the heroes having no way of knowing, in advance, that a tragic occurrence would bring out a special spark of courage in each of them.

While the frenzied rescue operation was being conducted in falling snow on the Potomac River, another one was taking place on the bridge itself for those motorists unfortunate enough to be there, stuck in traffic in the path of the roaring, descending airliner. Screams were heard amid the frightening sounds of metal and glass being destroyed as the immobile and helpless vehicles were struck by the aircraft.

"Some of the cars were flattened, some were without roofs, others were spun around," said the captain in charge of the rescue squads from the District of Columbia. "It looked like the plane just sliced the top of the cars." Rescuers trying to free bodies in the crushed vehicles were stymied by the fact that some of the cars had been flattened to door level.

As if the tragedy affecting the people on the bridge, and the one concerning the occupants of Flight 90 were not more than enough for the nation's Capitol area to cope with in one afternoon, another transit emergency occurred almost at the same time, within one-and-a-quarter miles of the crash site.

A crowded Metro subway train was derailed at rush hour in downtown Washington, D.C. Three passengers were killed, 25 others seriously injured and 1,320 people had to be evacuated from the system in the darkness and confusion that followed. Some of the rescue units from the District of Columbia and the surrounding communities were still enroute to the airplane crash site when the derailment took place. Several of these units were then redirected to the scene of the subway accident.

A Washington official stated that, unfortunately, the Capitol had been stricken that day with 3 coincidental disasters; the severe snowstorm, the jet airliner crash and the subway derailment. The National Transportation Safety Board agreed that, "the occurrence of two major accidents within a 30 minute period in the Washington Metropolitan area during weather conditions as they existed (that day), placed a severe burden on the emergency response capability of those jurisdictions required to respond to both accidents."

Despite the confusion, it did not take too long before an official casualty list for the Air Florida crash was issued. 73 of the occupants of the jet died from severe impact injuries while only one, with minor injuries, died from drowning. Five survived: four passengers and one flight attendant, all seriously injured. On the bridge, four commuters were killed and four injured.

(Note: It wasn't until 18 months after the crash, that a Coast Guard investigation concluded that the unknown hero who kept passing his lifeline to others, was not only the lone passenger who escaped serious bodily injury but was also the only victim who died of drowning. He was identified as a 46-year-old bank examiner from Atlanta, Georgia. The hero's mother was awarded the Coast Guard Gold Lifesaving Medal by President Reagan at an Oval Office ceremony at the White House in June, 1983.)

The five people rescued (as well as the one passenger who drowned) were all seated in the aft cabin area and managed to escape through the separation of the tail from the rest of the aircraft. This fact is not too surprising based upon the history of survivors of other aircraft accidents who were seated in the tail section of their airplane.

One survivor, late boarding the plane because he was busy parking his car, took a seat, "as far back as you could go, next to the galley." Although his boss, along on the same flight with 6 other colleagues, invited him to join them more towards the front of the airplane, he politely declined.

When the plane crashed, he found himself in the water. "I started losing my left shoe," he said. "I could feel it coming off and that was the most important thing in the world . . . I was thinking my foot would be cold." After being rescued, he learned that all 7 of his associates died in the crash.

There were also stories from relatives of some of those who did not survive. One woman recalled when her husband called her from the airport during the delay in departure. "He said, 'It's one helluva mess here. Sand trucks are everywhere. They're gonna board us, so once they get the runways cleared, we can go.'

"I said, 'Is it safe?' He said, 'I sure hope so, Hon, but they should know what they're doing.'"

One heartbreaking story came from the husband of a flight attendant who died in the crash. He flew up from their home in Florida when he heard the news. He journeyed out to the bridge to stare at the torn retaining wall and icy river. As he watched divers searching for bodies, he remarked that his wife had been a stewardess for Air Florida for two-and-a-half years and had only learned several weeks earlier that she was pregnant. He said, "We went for our first obstetrics exam and discussed whether she should continue to fly. So we said, 'Let's see how you like this trip and if you don't like it, we'll tell them you're pregnant and going off line.'"

He stated that he and his wife had never talked about the possibility of a crash. "I'd always kiss her goodbye," he said, "and tell her to come home safely. I always had the faith she would." He added determinedly, "I want her body out. I want it back in Florida. . . . My goal now is to get her out from under that frozen river and back to sunshine."

In all of this sadness, there was at least one note of relief. A tearful young lady, age 22, visited a church which had been set up as an information center for relatives of those believed to have been on the Air Florida airplane. She feared that her boy friend had taken that flight.

Uncertain as to the accuracy of the passenger list, she finally phoned him at home. "What the hell are you doing there?" she screamed when he answered. He had booked a different flight that cancelled because of the weather.

To conduct rescue and salvage operations and recover the bodies entombed in the aircraft (as well as those under the ice or at the bottom of the Potomac River), a total of 82 divers trained to dive in icy waters were brought in from various U.S. Navy, Army, and Coast Guard units. The divers would also be searching for the "black boxes" (flight data and cockpit voice recorders) which are important in helping determine the cause of an accident.

Their task turned out to be difficult since the water was quite murky; the underwater visibility was limited to only 8 inches. In addition, the wreckage was resting in freezing water, 25 to 30 feet deep, covered by floating segments of ice. The divers also had to contend with swift and shifting currents, and even with their insulated suits, could not remain in the water for more than one hour at a stretch.

It took one full week to recover both flight recorders. (They were found to be only superficially damaged with the "tape quality . . . good.") They were also successful in recovering all the victims—but that took longer. The last body (that of a two-month-old infant, whose father died in the crash and whose mother was one of the five survivors) was recovered ten days after the salvage operation began.

The National Transportation Safety Board (NTSB), which investigates every major accident in the United States, maintains a team of experts with different specialities on a 24-hour alert, so that they can respond immediately to news of a crash. The Safety Board claims that it has been able to find the "probable cause" of more than 95% of the crashes it has investigated and one of the reasons for this high ratio, besides the expertise of its staff, is the ability to get to the accident site quickly.

It is ironic that the Air Florida crash taking place "around the corner" of the headquarters of the NTSB, proved to be, because of weather and traffic conditions that day in Washington, difficult for the entire "go-team" to get to for several hours after being notified. Said the team's witness interviewer, "We got to this accident later than some of them we get to by flying."

The Safety Board conducted its usual thorough investigation. Seven months after the crash, the Safety Board issued its official report:

"The probable cause of this accident was the flight crew's failure to use engine anti-ice during ground operation and takeoff, their decision to take off with snow/ice on the airfoil surfaces of the aircraft, and the captain's failure to reject the takeoff during the early stage when his attention was called to anomalous engine instrument readings. Contributing to the accident were the prolonged ground delay between deicing and the receipt of ATC takeoff clearance during which the airplane was exposed to continual precipitation, the known inherent pitchup characteristics of the B-737 aircraft when the leading edge is contaminated with even small amounts of snow or ice, and the limited experience of the flight crew in jet transport winter operations."

Despite the inordinate number of errors, the Safety Board believed that this tragic accident would never have occurred if *only one* of the following actions had been taken:

1.  If the engine anti-ice system had only been turned on enabling the aircraft instruments to indicate an accurate reading of engine power. Or,

2.  If the captain had only issued a command to abort the takeoff. Or,

3.  If the flight crew, immediately after take-off, acting instinctively, had only poured on the power by pushing the throttles forward (like jamming the accelerator of a car down to the floorboards). That extra margin of thrust would have enabled the 737 jet to safely maintain its climb to an altitude where the snow and ice would have eventually shucked off.

What a pity—only one of any of those three "Ifs."

One can't help but imagine the relatives and friends of the 74 persons who perished, thinking, over and over again, sadly and regretfully, "If only . . ."

# Chapter 18

# Gunfire in the Cockpit

It would appear that the cause of practically every aircraft accident can be blamed on either a mechanical problem, adverse weather conditions or pilot error—or a combination of any of these factors. The year 1987 produced a rare exception.

On the afternoon of December 7, 1987, Flight 1771, a Pacific Southwest Airline four-engine British Aerospace jet airliner, took off from Los Angeles heading for San Francisco. Aboard the aircraft were 38 passengers and a crew of five.

One of the passengers making the trip was a disgruntled former employee of the airline who, after 14 years of service as a passenger agent, had been dismissed the month before on charges of stealing money from U.S. Air (the new parent company of Pacific Southwest). Also aboard this same flight, seated in another section, was his former supervisor who had recommended his dismissal.

While cruising at 22,000 feet, the ex-employee wrote a note to his former boss on an air-sickness bag which read, "Hi, Ray. I think its sort of ironical that we end up like this. I asked for some leniency for my family, remember. Well I got none and you'll get none." (It was not known whether the supervisor ever saw the note.)

When the jet airliner was about 175 miles out of Los Angeles, the pilot radioed a message to air controllers that he had an emergency in the form of gunfire in the passenger cabin and that he was setting his squawk box (transmitter identifying his aircraft to ground radar) on the emergency code. When the controller asked the pilot to repeat the message, the captain said, "I have an emergency—gunfire." That was the last communication from the captain.

Shortly afterwards, in deathly silence, the aircraft nosed over in an uncontrollable dive from a height of four miles which ended in a devastating plunge into a hillside. The impact was so severe that the largest piece of debris measured less than four feet long. Obviously, there were no survivors.

Authorities searching through the wreckage over the next several days actually found the charred air-sickness bag with the message written by the ex-employee. Also discovered among the bits of metal, fabric and other jagged scraps of debris was a .44 magnum pistol. (It was later learned that the disgruntled employee had borrowed a .44 magnum pistol along with a box of 12 shells from a fellow U.S. Air employee. It was assumed that, as a recognized airline employee, he was able to bypass the usual security screening for weapons.)

Besides locating these important clues, the search produced, of course, the Cockpit Voice Recorder which, although smashed like an accordion, safely cradled the vital tape which recorded cockpit sounds. Playing of the tape revealed that, in addition to the gunfire in the passenger cabin, there were 4 shots discharged in the cockpit area. Authorities have speculated that 1 bullet was fired at each of the three flight crew members with the last shot self-inflicted by the ex-employee.

(On February 25, 1988, the Secretary of Transportation informed Congress that regulations are being drafted to require the nation's largest airports to install security systems to screen the movements of airport and airline personnel. Magnetized identification cards would be issued to admit authorized employees through computerized gates which could be programmed to prevent ex-employees from gaining admission.)

# Chapter 19

# Bomb on Board

Consider the consequences of that last accident. Competent crew, airplane operating normally, weather conditions ideal—and yet, just one passenger venting his anger at another caused the destruction of the lives of scores of innocent people.

Acts of sabotage can be especially frustrating to air safety experts since they are the least expected. We now turn to a different type of sabotage, which officials are constantly on guard against: terrorists smuggling a bomb on board an airliner.

December 21, 1988 was a particularly happy time for many of the passengers of Pan Am Flight 103. Most of them were on their way to the United States to join their families for the Christmas holidays. Some were returning from business trips or diplomatic assignments, others were completing their vacations, military personnel were coming home on leave, and a contingent of thirty-five Syracuse University undergraduates were on a holiday break after spending a semester abroad.

Flight 103, scheduled to fly to J. F. Kennedy International Airport in New York and then on to Detroit, originated in Frankfurt, West Germany as a 727 three-engine jet airliner. Upon arrival in London, those passengers continuing on to the U.S. were informed that their luggage would automatically be transferred to the much larger four-engine 747. Other passengers, whose flight tickets and baggage originated in London, boarded the airplane to join them. Including the transferees from Frankfurt, the Pan Am 747 now carried 16 crew members and 243 passengers, 259 people in all.

The aircraft also carried a powerful plastic explosive concealed in a radio-cassette player packed in a suit-case stored in the baggage compartment.

This explosive is known as Semtex or C-4. As it is odorless it cannot be detected by trained dog sniffers. It also can't be readily identified by X-ray machines since the plastic, a putty-type substance resembling clay, can be molded into any shape. It can easily pass as a figurine, some wire, a piece of fabric or even as part of the luggage material itself.

Authorities believe that this particular bomb had two separate detonating devices to make sure the explosion would take place during flight. Damage to an aircraft while it is still on the ground, though severe, is not as disastrous as an explosion in the upper atmosphere where aircraft control cables can be severed and occupants and airplane subjected to explosive decompression from ruptured fuselage walls.

The initial detonator of this kind of bomb is a form of barometric instrument that is not activated until the airliner climbs to approximately 5000 feet. Regardless of how long a flight may be delayed on the ground, the bomb's miniature altimeter (an instrument that registers altitude) patiently remains dormant until the aircraft finally takes off. When the airplane reaches the target altitude, a timing device is then activated. If, for example, the timer is set for thirty minutes to give the aircraft enough time to reach the assigned cruising altitude, when this time has elapsed an electric circuit connected to a miniature battery closes, igniting a blasting cap embedded in the plastic explosive.

This was the scenario facing the occupants of the ill-fated flight. Unaware of the existence of the deadly instruments performing their fiendish tasks in the baggage hold beneath them, we can easily imagine the typical scenes on board—cockpit crew members scanning their instruments and monitoring their radios, flight attendants passing out magazines and preparing their beverage carts for their trips up and down the aisles and passengers dozing, reading and conversing with each other. While these routine activities were taking place, the initial detonator was about to achieve its goal of activating the second device to start its deadly time cycle.

The Pan American World Airways Flight 103, named Clipper Maid of the Seas, had taken off from Heathrow Airport at 6:25 PM, about twenty-five minutes behind schedule. The lateness was not surprising, considering the hectic activity at the airport at this time of year. The acceleration down the runway had been normal and the lift-off smooth as the Captain rotated the nose of the aircraft to the sky and began his climb. Communications with the controllers were pleasant as instructions were given and acknowledged by the crew. For veteran travelers, it was a routine transatlantic flight.

It was now fifty-two minutes after take-off. Flight 103 was twenty miles northwest of Carlisle and cruising at 31,000 feet over the town of Lockerbie, Scotland. The last radio contact had been made only two minutes before. Scottish radar observers following the flight were observing a normal radar image—until the second detonator circuit closed.

The battery set off the blasting cap in the high-performance plastic explosive located in the baggage compartment just aft and under the cockpit. The four-engine 747 airplane was blasted out of the sky. The force was so powerful that the entire front section of the 747 was torn completely apart from the rest of the aircraft.

There was no verbal communication from the Pan Am indicating a problem. There was no warning at all. The only indication of a possible disaster came from the radar operator at Prestwick, who noticed that the single radar image he was following had turned into multiple falling blips. Within minutes, even these signs disappeared from the screen.

The falling blips represented major sections of a torn-apart jetliner which, prior to the blast, had been flying close to 600 miles per hour. The explosive force at high altitude and high speed was so powerful that parts of the aircraft, together with its human cargo, continued to fall from the heavens, with devastating results, over an area more than ten miles long and eighty miles wide.

Unfortunately for the 4,000 unsuspecting residents of the small Scottish town of Lockerbie, located just north of and across the border of England, most of the heavy metal and burning fuel impacted on their homes and automobiles during dinnertime.

"There was a terrible explosion," said one Lockerbie resident. "The sky was raining fire." Said another, "The flames continued for a long time. We thought it was an earthquake."

Sections of fuselage caved in roofs and shattered windows, while heavy chunks of debris tore huge craters in the earth. Along with cascading metal, blazing fuel sprayed onto rows of houses. Many of these homes were set completely aflame in a huge fireball. Some residents, although badly injured and/or burned, managed to escape. Others didn't stand a chance.

Rescue operations were started immediately, with assistance from England as well as Central Scotland pouring into Lockerbie. Helicopters from RAF and North Sea support services were flown in to help firefighters and transport the injured to hospitals. The Dumfries Royal Infirmary, located 10 miles west of Lockerbie, received some survivors, but the majority had to be taken to major hospitals in Glasgow and Newcastle, about seventy miles away. However, some of the rescued were too far gone; eleven citizens of Lockerbie lost their lives.

As for Flight 103, there was no hope of anyone surviving. Helicopters and ground search crews were employed over the vast crash area to try to locate victims and fragments of wreckage. Remains of victims were transferred to a temporary mortuary set up near the town hall, while remnants of the shattered 747 (more than 10,000 items of debris) were brought to a warehouse near Lockerbie for later reassembling.

The extensive search included hundreds of volunteers along with the police, military, and special personnel assigned to the task. At times, more than 1,500 people participated in this painstaking endeavor.

Some areas were relatively easy to cover. Much of the wreckage was strewn around the streets of Lockerbie, including one engine lying just outside the village. The cockpit of the 747, still containing its crew members, rested conspicuously on a hill about three miles east of the village. Other bodies and large parts of the fuselage were lying in open fields, visible and accessible.

Other areas were difficult to explore. Since the winds at high altitude had dispersed the wreckage over a vast amount of territory, some of the ground to be covered consisted of dense forests and mountainous terrain. Authorities fear that, despite all efforts, some victims, as well as some of the 747 debris, may never be found.

(Intelligence officers believe that the terrorists had set the timing device to destroy the airliner over water. Had Pan Am 103 taken the usual route over the ocean, there would have been no wreckage to recover. Due to the prevailing winds at the time, the course of the airliner was changed to fly north over Scotland where the explosion took place.)

Family members awaiting the arrival of Flight 103 faced an evening of nightmare. Parents, relatives and friends, many laden with gifts, were escorted to a heavily guarded lounge when they inquired about Flight 103. There they huddled together in shock and grief as chaplains and Red Cross workers tried to offer comfort.

Many of them were bitter when they learned that passengers had not been made aware of a memo circulated to U.S. Embassy employees. This memo alerted them to a telephone call received by the Embassy in Finland about three weeks prior to the crash, which stated that a bombing attempt would be made within two weeks against a Pan Am airliner flying from Frankfurt to the United States. Although investigation by the Finnish authorities later disclosed that the telephone call was a hoax (and, furthermore, the two-week threat period had expired without incident), relatives still felt that passengers should have known of the existence of the threat; then at least they would have had the choice of cancelling their reservations on Flight 103 or continuing their plans as scheduled. The response by U.S. spokesmen, as well as airline personnel, is that such calls are received every day and, almost always, turn out to be false. "If we shut down airlines every time there is a threat, we will allow terrorists to disrupt the economy, not only of this nation but of the world," said the U.S. Secretary of Transportation. Despite these statements, some families filed multimillion dollar suits against Pan Am and the companies that search cargo and baggage, claiming that adequate security was not provided to protect passengers against a terrorist attack.

(Newspapers later reported that the British Ministry of Transport had also issued unpublicized alerts before the Pan Am 103 disaster warning British airlines, and other carriers, of the new type of terrorist bomb that could be hidden in a radio-cassette player. Pan Am claimed that it did not receive the British bulletins until several weeks after the bombing.)

Actually, after the Finnish telephone threat, security had been tightened at both Frankfurt and Heathrow airports. However, it is almost impossible to achieve absolute security given the present detection equipment available to airlines. Despite the imposition of stiff penalties against United States airlines whose security procedures are deemed to be inadequate, even the Federal Aviation Administration admits that almost 10% of simulated weapons manage to get by checkpoints during tests conducted by the Agency.

Although hand weapons smuggled on board are dangerous, explosives are even more so. After the Pan Am disaster, the FAA issued orders that would apply to all U.S. carriers operating at airports in Western Europe and the Middle East. Baggage must be physically inspected as well as X-rayed and must be loaded on the same plane as the passenger. In addition, some passengers, based on the judgment of trained security guards, will be subjected to a body search. Ground service employees, who have easy access to the aircraft, will also undergo more stringent scrutiny.

The bombing of Flight 103 turned out to be the worst airline disaster in British history. Her Majesty the Queen of England was appalled when she heard of the tragedy and asked to be kept informed. Both Prince Andrew and Prime Minister Margaret Thatcher visited Lockerbie while smoke still hung over the ruins. Viewing the carnage, Mrs. Thatcher stated that she was "shocked by this terrible disaster" and expressed her sympathies to the bereaved families. United States President Ronald Reagan extended his condolences and said, "Our hearts go out to you on this tragic occasion, which marred what should have been a season of joy."

The Prime Minister, along with other British Government officials and the American Ambassador, attended a memorial service held two weeks after the crash. Also in attendance were Lockerbie townspeople, almost 200 American relatives flown in by Pan American, and more than 100 Pan Am employees. It was an emotional experience for everyone memorializing the 270 randomly chosen, unfortunate innocents.

The large number of fatalities from the crash of Flight 103 mobilized investigating resources in Great Britain, Scotland and the United States. Initially, it was thought possible that structural failure of the 747 or a midair collision with a military aircraft at high altitude might have caused the destruction of the airplane. However, after examining the remains of the shattered radio-cassette player and fragments from a metal luggage bin, British ordnance technicians in Kent soon concluded what had been suspected all along—the culprit was a high-performance plastic explosive.

Since terrorists appeared to be most likely responsible for the bombing, most of the world's intelligence community was galvanized into action. Dozens of nations offered their services, including the French, Germans, Israelis, Italians and Swiss. Even the Soviet Union and the PLO announced that they would try to be of assistance.

Great Britain has taken the lead in this inquiry. Scotland Yard is coordinating efforts with the U.S. Federal Bureau of Investigation (FBI), the Central Intelligence Agency (CIA) and the National Security Agency (NSA). Both nations have extensive experience dealing with terrorist organizations. Sharing information that heretofore has been zealously guarded will be extremely valuable to the inquiry.

In addition to intelligence agencies, the aviation accident authorities will also be joining forces. The National Transportation Safety Board of the U.S. will be cooperating with the British Department of Transport Air Accidents Investigation Branch.

A painstaking amount of detailed checking for clues will have to take place. In addition to interrogating all airline employees who had any contact with Flight 103, including maintenance personnel, counter attendants, and baggage and cargo handlers, the list of passengers and crew members will be examined and every name turned over to investigators. Recent travel itineraries will be looked at, relatives queried, names of friends obtained, and the recipients of expected packages called upon. Every possible lead will be followed up.

The investigating teams will also try to determine if there is any link to other airline bombings by studying some of these other recent acts of terrorism:

November 29, 1987—A South Korean jetliner was blown up over the Thailand-Burma border, killing all 115 people aboard. The government of South Korea claimed North Koreans were responsible.

April 2, 1986—A bomb placed under a seat exploded on a TWA flight from Rome to Athens. Although the aircraft was able to land, four people were killed and nine were injured. Police stated the bomb was left under the seat by a woman with a Lebanese passport who got off the plane in Rome.

June 22, 1985—A 747 Air India flight from Toronto to Bombay, by way of London, with 329 people aboard, was blown apart off the coast of Ireland. Anonymous callers said the Sikhs were responsible. (This disaster resembled that of Pan Am 103 in that both aircraft were 747's carrying hundreds of passengers, and, in both cases, the recovered cockpit voice recorders revealed normal flight deck conversations taking place until abruptly cut off by a similar unidentified noise at the end of the tape.)

In addition to these incidents, another bombing disaster was averted in 1985 when an El Al security guard became suspicious of an Irish woman in possession of a heavy carry-on bag who was about to board an airliner scheduled to fly from London to Tel Aviv. A search of her luggage disclosed three pounds of the plastic explosive, Semtex, which had been planted without her knowledge in a false bottom of her suitcase by her Jordanian-born boyfriend.

Authorities feel that if they can discover some common thread in these incidents—the same explosives or terrorist groups involved—it might help them in their current investigation.

The inquiry on the Pan Am Flight 103 bombing has involved more people and resources than any other air disaster in recent history. Authorities are confident that they will determine exactly how and when the explosive was placed and which group or groups were responsible.

In the meantime, what can be done to prevent other bombings? Up to now, the following techniques have been employed for international travelers:

1. Dog sniffers
2. Metal detector equipment
3. X-raying of all baggage (both carry-on and that placed in the cargo hold)

4. Physical inspection of all baggage
5. Requirements that passengers and baggage travel on the same airplane
6. Body searches where deemed warranted
7. Additional training for security guards
8. Strict screening and indentification of maintenance personnel
9. Interrogation of passengers. Examples: "Did you pack your own bag? Did anyone else have access to it? Who did you visit? Where did you stay? Did you receive a gift? Are you carrying a package for anyone? Whose car brought you to the airport?"

Most of these measures have long been used by Israel, a prime target for terrorists. Israeli security agents go even further to prevent acts of sabotage. They are trained to pick up clues by looking into the eyes of the passengers they are questioning and noticing how they respond. In addition, at times, they place luggage in pressurization chambers to prematurely detonate bombs using barometer-type devices. Counterterrorism experts say that Israel has the best security system in the world.

Despite the present methods to prevent explosives from being smuggled aboard airplanes, there is, as we have seen, one major gap in our security system. Plastic explosives, composed of a putty-like odorless substance that can be molded into any shape, easily escape both our present vapor sniffers and X-ray detection devices.

At the urging of the FAA, the remaining loopholes are about to be closed. With FAA financial assistance, two new technological detection devices are being developed. One is called Thermal Neutron Activation, or TNA. Baggage, while passing through a chamber, is bombarded with neutrons which produce gamma rays that are analyzed by a computer for the existence of explosives. The technology has been so effective in detecting explosive material that the FAA has ordered 6 units to be placed at U.S. airports. The cost of each unit is approximately $1 million but is expected to decline as additional equipment is ordered. All airlines will be required to have then installed after the initial devices have been satisfactorily employed. It is expected that at least 50 units will be in operation at international airports by the end of 1990.

Because of the hazard of radiation, TNA cannot be used on humans. The other detector device does not use radiation and, therefore, can be used. It is an electronic sniffer. Passengers passing through a small booth receive a five-second blast of warm air. Electronic equipment samples the air for fumes that can indicate the presence of explosives. The initial units of this electronic detection device will also soon be in operation at U.S. airports.

Thus, as a result of the disaster over Lockerbie, all airlines will soon be equipped with sophisticated detection devices and search procedures that will make it much more difficult for explosives to be carried on board any commercial aircraft. Let us hope that the reaction of the world to the violent ending of Pan Am Flight 103, as tragic as it was for the occupants and families of the victims, will serve as a catalyst for the air traffic industry to make future flights safe from all acts of terrorism.

# Chapter 20

# Wrong Engine Shut Down

"This is Midland 92. I have . . . engine failure in my starboard [right] engine and smoke in the cockpit . . . I am diverting to East Midlands airport."

This was the message sent by the English Captain of a Boeing 737-400 twin-engine jet less than 3 weeks after the Lockerbie disaster. The aircraft, British Midland's Flight BD 92, carrying 118 passengers and 8 crew members, had taken off from Heathrow at 7:52 PM on January 8, 1989, on the 70 minute air shuttle flight to Belfast. This flight is especially popular on Sunday evenings with British businessmen and government officials commuting back to Northern Ireland after having spent weekends in England.

The radio transmission to the ground controller had been stated in a matter-of-fact, calm tone. After all, the sophisticated airplane, now cruising at 30,000 feet, was practically brand new, having been recently shipped from the Boeing factory. It was powered by two modern jet engines, each one powerful enough to propel the jetliner safely to its destination. In addition, the pilots, as part of their training twice each year, have had considerable practice maneuvering the 737 with only one engine operating.

After consulting by radio with the British Midland's operating staff at Castle Donington, the captain notified air traffic control that he was shutting down the starboard (right) engine. He then made an announcement over the public address system, in reassuring tones, that he had shut down the right engine and the flight would be diverting to another airport. This announcement was received by surprise since some passengers had noticed, in addition to loud bangs on the *left* side, sparks and "bright lights" coming from the port (left) engine underneath the wings. No one mentioned the apparent discrepancy to the flight attendants, who were hurriedly collecting the trays of uneaten food that had just been distributed. No one attempted to question the authoritative, reassuring statement of the captain.

One perplexed passenger later remarked, "I looked over at the left wing and saw flames and flashes coming from the engine. Cabin staff stopped serving meals and went to the front of the plane as the pilot told us not to be unduly concerned." He added, "The captain apologized and said there was a fire and that we were diverting to East Midlands airport. Two girls started to scream but people were generally very calm."

The captain, despite the shutdown of the right engine, appeared to be in complete control as he throttled back the left engine in preparation for a descent. Any possible problem with the left engine was no longer apparent to the passengers as the loud bangs, flashes, and "bright lights" disappeared with the easing of power. Some of them now turned their thoughts to the difficulty they would have getting to Belfast once the plane arrived at the diverted airport.

While adjusting the controls of the 727 to compensate for the absence of power in the right engine, the veteran captain, with over 25 years of experience with British Midland, responded to the controller's radio query as to how many "souls" he had on board by transmitting "117 plus 1 [infant] and 8 crew."

Instructions were then given by the Manchester air traffic control to descend from 30,000 feet in a wide, gentle bank over Nottingham for a direct heading to the East Midlands Airport. Shortly afterwards, the co-pilot alerted the airport controllers that Midlands flight 92 only had one engine operating and was coming in on "a wing and a prayer."

The final radio message from the Manchester controller before handing the airliner over to the control tower was, "Descend to 7,000 feet. Bank right . . . I hope you make it." The captain replied, "Roger, so do I."

The aircraft continued its descent to about 1,500 feet above the ground and within 5 miles of the runway. The landing gear was then lowered, and the captain slowed the descent by raising the nose of the aircraft. With the airspeed down to 120 miles per hour in preparation for landing, the captain now advanced the throttle to his left engine to carry the airplane to the runway. But the power was not there. The only result of trying to apply power to the port engine was the initiation of a series of backfiring from that engine, which now started to break up and shed debris along its flight path.

Startled by the disintegration of the engine they had counted on, the co-pilot shouted, ". . . there's the other one." The captain, who at the time was communicating with the control tower, remarked, "Did you copy [hear] that? You know we have got trouble."

A passenger recalled, "There was a large chunking noise. It was obvious there was some mechanical fault." An observer on the ground reported, ". . . something was very badly wrong. The plane was vibrating very badly. It seems the whole thing was shuddering." Another observer stated that flames began to shoot from the left engine, which was making deep, loud, "Booming noises."

The captain, trying to coax some power from his reluctant port engine, was on his final approach over the village of Kegworth, only 60 seconds from the edge of the runway. With the left engine unable to produce sufficient thrust and no time to try to restart the right engine, the captain raised the nose of the airliner 45 degrees in an attempt to gain altitude and stretch his glide.

Although both its engines were now crippled, Flight 92 had managed to reach the point where the runway was in sight. But there was one remaining obstacle to overcome—a 30-foot embankment just above the M1 motorway, England's main north-south highway. A Kegworth resident living near the airport witnessed the scene. He said, "I heard a plane coming toward the motorway. It was making a tremendous noise with an engine appearing to be backfiring. It was obviously in trouble.''

The crew was helpless. The embankment barrier between them and the runway appeared to grow in height as the shuddering, rumbling and clanking Boeing 737, with its flaming left engine back-firing and spitting metal fragments, slowly but inexorably drifted towards the ground in a 45 degree nose-up position. Watching the airplane settle, another observer said, ''How the pilot kept it up, I don't know. It came in very low, going boom-bang-boom, the queerest, scary noise, like a car with its big end gone. It made you go cold. It was all lit up, they had every light on they could, and it seemed to be glowing as well. But all the time you could see they were desperately trying to get the nose up.''

Rapidly approaching the looming, insurmountable M1 embankment, the captain's final announcement over the public address system alerted the flight attendants and passengers to prepare for a crash landing. One survivor said, ''It seemed like only five seconds later we were down. The lights went out and we seemed to be stuck at an angle of 45 degrees. People were moaning. My main worry was that there would be a fire.''

Another survivor concurred, ''The pilot came on and told us to prepare for a crashing landing. Five seconds later we were on the ground. We didn't have time. There was a big thump but there was no panic.''

The 737, with its nose up, hit the ground on its belly just before the embankment, only 200 yards short of the airport perimeter. It then tore its way through the trees that line the steep embankment, continued across the highway crumpling the central crash barriers, before hurtling into the opposite embankment, where it came to rest, broken into three sections. A small fire then ignited near the main fuel tanks.

Recalled a passenger, ''I had my head forward in the crash position. Suddenly the seats came up behind me and I was lodged. The plane actually split behind me. The lights were out and it was total chaos. There were flames at the front and I thought, 'That's it, we're going to get burned.' Then there was foam. That was the best part of it.'' He was rescued by firemen, who appeared on the scene almost immediately. They played a vital role in keeping the casualty list from mounting.

Another survivor also had praise for the fire rescuers. "My wife couldn't move. She was pinned down by some metal and a piece of plastic up against her throat. It seemed only a minute later and someone was crashing the windows in over the wing. The man in [aisle seat] 15C helped pull the debris off us. He helped to get my wife out and through the window exit, then he helped me.

"It was remarkable—thirty seconds after I was out, the plane was covered in a gush of foam. I remember my suit was soaked and it was slippery underfoot. Only sixty or ninety seconds after we hit the ground, the firemen were there. They were everywhere."

One of the first people on the scene said, "The seats had all gone forward towards the cockpit end and the people had all been shunted together. There were people in a sort of concertina. You had to take the seats from the back of the people in front to get them out.

"All of them seemed to have smashed ankles and legs, and with a lot, their faces were badly damaged too. It seems a miracle that anyone got out alive though, and I can't believe that it all didn't go up in flames. It was horrendous."

Another rescuer said, "I helped to get five people out, including the pilot. When I went into the front section of the aircraft, it was completely ripped away. There were seats and bodies piled up everywhere. The injured were moaning, but nobody seemed to be panicking.

"The pilot was trapped in the cockpit. We had to cut away part of the seat and bulkhead to free him. He was moaning as he was carried down the hillside."

Firemen, policemen, village residents, medical personnel, and more than thirty ambulance crews from throughout the Midlands helped assist passengers trapped in the wreckage. The leader of a medical flying squad from Leicester Royal Infirmary stated, "It was absolute carnage. Access within the aircraft was extremely difficult. Chairs and internal partitions had collapsed. I am surprised so many people have survived."

Unable to maneuver inside the shattered airliner, doctors had to wait hours for firemen to cut through the wreckage before they could tend to trapped passengers. In a few instances, amputations of limbs had to be performed to free victims. Said a member of that medical team, "Given the circumstances, we had no choice."

For the most part, survivors remained calm. An emergency official recalled, "There was no panic. Passengers cooperated totally with the rescue services. We were requiring them to stay very still and not be frightened by the metal-cutting equipment working near them." The chief ambulance officer also noted that the morale of the trapped passengers remained high. He remarked, "Some of the casualties were conscious and talking. There was very good spirit inside the plane, especially with the people we could not actually get to but could talk to."

The use of seat belts by passengers, together with the partially controlled descent of the nose-up aircraft impacting on the underside of the main fuselage, kept the accident from becoming a total disaster. Initial reports indicated that approximately one third of the occupants perished immediately, another third sustained serious injuries, and the remaining third received minor injuries, with some actually walking away from the wreckage unscathed.

The captain, who suffered a fractured spine and other injuries, also helped reduce casualties by guiding the aircraft away from residential areas nearby. The low-flying jet airliner could easily have turned Kegworth into another Lockerbie disaster. A special service was held by the Rector of Kegworth to give thanks that the residents who live in the East Midlands flight path had been spared.

Motorists traveling that evening on the M1 were also fortunate to have avoided colliding with the fallen 737 aircraft and its debris. "There was a gap of about 200 yards between the traffic on both sides of the motorway at that time," a police inspector said. "That's quite extraordinary for a Sunday evening with people returning home after the weekend."

Two chaplains, who less than a month ago had spent innumerable emotional hours comforting the families of those killed in the Lockerbie air disaster, were now called upon again to offer their services at Heathrow Airport. This time, some relatives receiving heartening news. "We were able to tell one young lad that his brother had survived the crash. He was found in some trees nearby with two broken legs."

The other chaplain added, "Another person went home happy after learning that the person he was worried about had not been on the plane. He had a standby ticket but he had not boarded the flight."

The Prime Minister found herself paying another visit to an accident scene. Mrs. Thatcher said, "We couldn't believe it, coming so soon after the crash at Lockerbie." She continued, "The action of the fire services was marvelous. If they had not kept the fire down, the emergency services would not have been able to get people out." She said with pride, "People who were trapped were very, very brave indeed. Doctors and nurses were in considerable danger themselves and no one can surpass our services."

The Queen also sent a message to the families of those who died. She said, "I was deeply shocked by last night's news of a second major air disaster, and relieved to learn that there were, on this occasion, many survivors from the crashed aircraft."

Although most of the 126 passengers and crew survived the crash, 46 people were fatally injured. Included in the death toll were six British service personnel, three sisters flying home to an Irish village to attend their father's funeral, and the mother of the lone infant passenger.

As expected, the Transport Department's Air Accidents Investigation Bureau, although already immersed in examining the Pan Am 103 wreckage, immediately dispatched a team of nine inspectors to the new accident scene. Here the investigation would concentrate on the cause of the failure of the engines of Flight 92. Assisting them would be technical experts from Boeing and General Electric.

The engines of the Boeing 737-400 airplane, designated CFM-56, are manufactured by CFM International, a joint venture of the U.S. General Electric Company and Snecma, a French concern. These engines have been in use on the 737-300, the Airbus A320 jetliner, and other military aircraft since 1981. They are the world's biggest selling civil aircraft engines and are considered very reliable; up to now, only one other engine-related accident has involved that model. This incident occurred in the United States in May, 1988, when a 737-300 lost power in both engines after taking off in heavy rain from an airport in New Orleans.

Though it managed to make a safe landing in a field, after that accident the Federal Aviation Administration directed airlines to increase the speed of their engines in heavy precipitation. In response to a question as to whether the British Midland Flight 92 had also been vulnerable to losing both engines, a FAA spokesman in Washington said, "We did put out a directive saying that these items could lose power when in heavy rain, and if you get into heavy rain, use . . . particular power settings . . . but there was no heavy rain on the night of that accident."

The Flight 92 investigators were seeking the answer to many questions:

1) Why did the left engine fail?
2) Why was the right engine shut down?
   A. Was this due to a faulty instrument or crossed wiring that gave an erroneous identification of the engine having a problem?
   B. Was it pilot error?

As for the loss of power in the left engine, the experts will be looking for a possible failure in the jet engine's ignition system or for the usual signs of metal failure, such as a fan blade becoming detached or a rupture at the combusion end of the power plant. Other possibilities to consider will be oil caps improperly fitted during servicing causing oil leaks, or a faulty batch of materials used in the manufacture of the engine.

The failure of the left engine was at least clear cut in that the cause was mechanical. The shut down of the right engine presented a more complicated puzzle. Did the captain receive a false reading indicating a problem?

On the advice of the investigators, the Civil Aviation Authority grounded British Boeing 737's while checks were completed on their engine fire overheat and vibration circuitry to make sure the sensing equipment differentiated accurately between the aircraft's right and left engines. This order affected nine British airlines with a total of 28 aircraft. This was followed by another order calling for urgent checks on all other British aircraft using CFM-56 engines.

The United States followed Britain's lead in ordering inspection of relatively new Boeing airplanes to make sure their fire warning and monitoring circuits were not flawed. The Boeing company later announced that no wiring discrepancies were found in the engine-warning systems of either the 737-300 or the 737-400 models. However, they did report wiring errors in the cargo fire extinguisher systems in eight twin-engine 757 models, errors which would have activated fire controls in the wrong cargo section. After this discovery, the company announced that it had altered the electrical and plumbing connections in the fire systems so that it was no longer possible to install them incorrectly. (Boeing, the largest manufacturer of civil aircraft in the world, currently has a backlog of orders for over 1,100 new aircraft. The company states that it is determined to maintain its excellent reputation by improving its quality control to prevent future miswiring.)

Since inspections of all installed CFM-56 engines found no clues to the failure of Flight 92's engines, and wiring errors were not discovered on any Boeing 737-400 models, officials have announced that, despite public impatience, the investigation would take longer than expected. The 45 tons of wreckage were removed to the accident investigation headquarters in Farnborough for reassembling and examination. The engines were shipped to Snecma in France for inspection.

Until the investigation is completed, the possibility of human error, although highly unlikely considering the experience of both the pilot and co-pilot, cannot be ruled out. However, when some officials advanced this theory, others defended the two cockpit crewmembers. The chairman of British Midland said, "I understand why quick judgements are desirable, but wrong information is worse than no information at all and the swing from hero to anti-hero which we have seen over the last 24 hours is wholly discreditable." The hospital doctor in charge of the intensive care unit also said, "[The captain] is an extremely intelligent and professional man and from my personal assessment I find it hard to believe that two experienced and competent pilots could make such an error."

Finally, Mrs. Thatcher, highly critical of all the guesswork and wild speculation over the cause of the crash, best summed up the feelings of a majority of British citizens when she told the Commons, "It would be far better if everyone accepted that the best course after a terrible tragedy of this kind is that those charged with the responsibility of that task are able to get on with that task and that nothing should be said until it is properly and truly completed."

# Epilogue

Now you know just how and why these accidents happened. I'm sure you'll agree that, despite the initial intensive coverage by the news media, the circumstances that ultimately led to each crash did not see the light of day until long after the accident occurred. And, even then, the inside story only appeared in professional aircraft journals, along with a brief article or two on the back pages of a few newspapers.

However, we suspect that after you have learned how easy it is for unanticipated events to trigger forces that can lead to an accident, you may have become a little apprehensive about taking your next airplane trip.

You may tell yourself that what you didn't know before, was not too frightening. Now, you may begin to question every announcement that the crew makes and be suspicious of every bump that the flight takes. You may even go so far as to interpret every strange creak, groan and sound you hear as a warning that your airplane is about to come apart in midair.

Therefore, we would be remiss if we ended this series of accidents on this uneasy note. We feel it is important, despite what you have read, that you understand how safe flying really is, so that your confidence in traveling by air is restored.

To start with, all of us accept the fact that regardless of what resources are brought to bear, there is little chance that all air crashes can be eliminated. As long as aircraft are controlled by crew members and controllers, no matter how well-trained, there is bound to be an occasional error in judgement. In addition to the human factor, our safety is also riding on mechanical and technological equipment which, despite rigid inspections and regularly scheduled preventive maintenance, sometimes does fail at the most inopportune time.

Yet, at the very least, can something be done to reduce the frequency of aircraft accidents? Absolutely. Something is being done. Surprisingly, as incongruous as it may seem, every time an accident occurs, our flight safety goal moves another step forward.

As you know by now, every accident of any consequence involving airlines, as well as many small general aircraft, triggers an investigation. In Great Britian, the inquiry is conducted by the Department of Transport Air Accidents Investigation Branch. Its counterpart in the United States is the National Transportation Safety Board. Both Agencies have professional teams available to immediately fly to the accident site and set up working groups to investigate weather, air traffic control and aircraft maintenance records. The team will arrange for public hearings, interrogate crewmembers, witnesses and possible survivors. They will also review tapes from recovered "black boxes," examine the damaged airplane (even to the extent of arranging to have totally destroyed aircraft reassembled by hand), authorize air and stress tests under simulated conditions and, if necessary, call on outside air experts— everything and anything to ultimately determine the exact cause of the accident. Upon the completion of a painstakingly thorough investigation, a written report is issued reviewing the circumstances that led up to the accident and coming to an official "Conclusion" as to just how and why the accident occurred.

The air accident investigation agencies will then recommend that specific orders be issued to either modify the procedures to be followed by air personnel and/or require the modification of aircraft equipment, depending upon the conclusions of the investigation.

Thus, in a strange sort of way, as unwelcome as it is in its arrival, each accident serves a purpose. It is almost the last of its kind. It should be somewhat reassuring to know that when the recommendations are enacted, the chances of a similar accident under similar circumstances occurring is indeed rare.

Which leads us then to a very important question that you may ask—The next time I fly, what are the chances of *my* becoming an accident statistic? To give you the answer, it is necessary to examine the accident records involving all types of aircraft and compare them with travel by ground transportation.

Since air travel in the United States is so much more extensive than in any other country, we will examine the history of U.S. airplane accidents in order to furnish more meaningful statistics.

There are many different types of U.S. aircraft carving out a portion of airspace at any time of the day or night. In addition to the large commercial airliners, there are corporate aircraft, small air-taxi and commuter airplanes, helicopters, and a horde of small general aviation planes (more than 200,000 of them) flown by private pilots for pleasure and/or business. We also have gliders, ultralights, balloons and even an occasional blimp.

Within the last 10 years, accidents involving all aircraft resulted in causing an average of approximately 1,400 deaths each year. The small general aviation airplanes, because of their considerable numbers, have been responsible for most of these fatalities.

1,400 deaths a year may seem high—until you compare it to the number of lives lost from accidents involving ground transportation. Approximately 50,000 deaths occur in the United States every year as a result of highway accidents. Yet, that frightening figure has not lessened the American love affair with the automobile. Nor, obviously, has it instilled in automobile travelers the same kind of fear that exists for many people contemplating a trip by airplane.

Now, let us examine the recent statistics on accidents only involving the large scheduled U.S. airliners (excluding the small air-taxi and commuter services) to try to get a perspective on the chances of a passenger boarding one of these commercial aircraft ending up as a fatality. The figures may very well surprise—and reassure—you.

Today, approximately 15,000 scheduled commercial passenger

aircraft take off from some airfield in the United States every single day of the year. This amounts to one airliner lifting off into the sky every six seconds round the clock. No wonder that scheduled airline traffic has reached and passed the figure of 400 million passengers annually.

Based on these astronomical numbers, when you consider the shrieking publicity generated by the newspapers, radio and television stations every time an airliner crashes, you would have every reason to expect that fatalities would be quite high. Let's take a careful look at the actual figures.

Starting from January 1976 through December 1988, the number of fatal aircraft accidents from these flights averaged—and this may be hard to believe—only three per year.

In that same period of time, with several hundreds of millions of passengers flying every year, fatalities averaged about 135 per year. (Less than one death recorded for every two million passengers.)

The decade of the 1980's was an exceptional period. The years 1980 and 1986 both had perfect scores in that there was not a single fatality in each of those two years. And, 1981 almost matched that safety record. Although 1981 experienced four fatal accidents, each one resulted in only one death in the following manner:

- An aircraft mechanic fatally injured while servicing the nose gear doors.
- A ground crewman run over during pushback.
- A passenger falling from the boarding ramp.
- A flight attendant crushed by a galley personnel lift door.

As you can see, 1980, 1981 and 1986 turned out to be extremely safe years for passengers traveling by scheduled U.S. air carriers. More than 1 billion passengers over a three-year period with only one passenger fatality—and that one from falling off a boarding ramp. A safety record that borders on the incredible.

Now ask yourself—do you feel a little more secure traveling on a commercial airliner? After all, you have now read about many of the steps that have or are being taken to prevent accidents and protect passengers when they do occur. You undoubtedly have also been impressed by the infinitesimal number of fatalities considering the vast number of passengers carried by the scheduled airlines each year. We feel sure that it has to have some beneficial effect on your conception of the dangers connected with air travel.

If this is so, we have a suggestion to make concerning the very next flight that you take. After you are seated and informed that the aircraft is shortly about to begin its taxi, fasten your seatbelt, place both feet firmly on the floor, close your eyes for about 10 seconds—and concentrate on the odds that are so heavily in your favor of you safely reaching your destination (2,000,000 to 1). Then, open your eyes, sit back, relax—and enjoy your flight.